Springer Theses

Recognizing Outstanding Ph.D. Research

Aims and Scope

The series "Springer Theses" brings together a selection of the very best Ph.D. theses from around the world and across the physical sciences. Nominated and endorsed by two recognized specialists, each published volume has been selected for its scientific excellence and the high impact of its contents for the pertinent field of research. For greater accessibility to non-specialists, the published versions include an extended introduction, as well as a foreword by the student's supervisor explaining the special relevance of the work for the field. As a whole, the series will provide a valuable resource both for newcomers to the research fields described, and for other scientists seeking detailed background information on special questions. Finally, it provides an accredited documentation of the valuable contributions made by today's younger generation of scientists.

Theses are accepted into the series by invited nomination only and must fulfill all of the following criteria

- They must be written in good English.
- The topic should fall within the confines of Chemistry, Physics, Earth Sciences, Engineering and related interdisciplinary fields such as Materials, Nanoscience, Chemical Engineering, Complex Systems and Biophysics.
- The work reported in the thesis must represent a significant scientific advance.
- If the thesis includes previously published material, permission to reproduce this must be gained from the respective copyright holder.
- They must have been examined and passed during the 12 months prior to nomination.
- Each thesis should include a foreword by the supervisor outlining the significance of its content.
- The theses should have a clearly defined structure including an introduction accessible to scientists not expert in that particular field.

More information about this series at http://www.springer.com/series/8790

Jian Yang

Study on Ground Moving Target Indication and Imaging Technique of Airborne SAR

Doctoral Thesis accepted by
Chinese Academy of Sciences, Beijing, China

Springer

Author
Dr. Jian Yang
China Academy of Launch Vehicle Technology
Beijing
China

Supervisor
Prof. Yanfei Wang
Institute of Electronics
Chinese Academy of Sciences
Beijing
China

ISSN 2190-5053 ISSN 2190-5061 (electronic)
Springer Theses
ISBN 978-981-10-3074-1 ISBN 978-981-10-3075-8 (eBook)
DOI 10.1007/978-981-10-3075-8

Library of Congress Control Number: 2016961666

Printed on acid-free paper

This Springer imprint is published by Springer Nature
The registered company is Springer Nature Singapore Pte Ltd.
The registered company address is: 152 Beach Road, #22-06/08 Gateway East, Singapore 189721, Singapore

Supervisor's Foreword

GMTI and GMTIm are important functions of airborne SAR systems, which endow the SAR system with the abilities of indication, parameter estimation, relocation and focusing of moving targets, as well as high-resolution imaging of stationary scenes. With these advantages, the moving target processing functions of SAR systems are widely used in both military and civilian applications, and are becoming hot research issues recent years.

With the development of signal processing theory and electronic technique, the airborne SAR system has also exploited several new operating modes. Besides the classical stripmap SAR/GMTI mode, the WAS-GMTI mode and FMCW SAR system are two representative new operating modes, which are drawing more attention of researchers. Based on the deep analysis of the signal model of the moving target, Dr. Yang focuses on the theories and applications of GMTI and GMTIm algorithms in SAR/GMTI mode, WAS-GMTI mode and FMCW SAR system, and also studies the corresponding key techniques of these operating modes in this thesis. In addition, the non-ideal errors that deteriorate the performance of GMTIm algorithm in real SAR data processing are discussed, and the compensation methods are provided.

This thesis is highly original and its results not only significantly deepen the theory of moving target processing technique but also have great practical values in real SAR data processing. I feel very lucky to have Yang as one of my Ph. D. students because he is a talented and diligent student. His work during the Ph. D. study promoted the moving target processing performance of the airborne SAR systems in our lab, and won him the President's Special Award of Chinese Academy of Sciences. This thesis improves the GMTI and GMTIm theories of airborne SAR, and paved a way for further development of real data moving target processing.

September 2016 Prof. Yanfei Wang

Preface

GMTI and GMTIm are important functions, which endow the SAR system with the abilities of the indication, parameter estimation, relocation, and focusing of moving targets, as well as high-resolution imaging of stationary scenes. With these advantages, the moving target processing functions of SAR systems are widely used in both military and civilian applications, and are becoming hot research issues in recent years.

With the development of signal processing theory and electronic technique, the airborne SAR system has also exploited several new operating modes. Besides the classical stripmap SAR/GMTI mode, the WAS-GMTI mode and FMCW SAR system are two representative new operating modes, which are drawing increasingly more attention of researchers. Based on the deep analysis of the signal model of the moving target, this book focuses on the theories and applications of GMTI and GMTIm algorithms in SAR/GMTI mode, WAS-GMTI mode, and FMCW SAR system, and also studies the corresponding key techniques of these operating modes. In addition, the nonideal errors that deteriorate the performance of GMTIm algorithm in real SAR data processing are discussed, and the compensation methods are provided. The main work and innovations of this book are as follows:

(a) The echo signal model of the ground moving target in stripmap SAR/GMTI mode is established. By deeply analyzing the relationship between the motions and Doppler phases of the targets with fast radial velocities, the effects of Doppler ambiguity and third-order phase error are studied, and a new classify scheme of the moving targets is presented by concerning the Doppler ambiguity. Through the Doppler centroid estimation algorithm based on curve fitting, the Doppler centroid error of the clutter can be estimated and corrected, which is the foundation of the following processing steps. With the two-step GMTI algorithm, fast-moving targets that submerged by the clutter can be indicated and extracted. Using the GMTIm algorithm based on Hough transform and third-order PFT, the Doppler ambiguity of the moving target can be accurately estimated, and the moving target can be correctly focused and relocated. Besides the advantages of indicating and imaging fast-moving targets, the proposed algorithms are also effective in slow-moving target processing, which makes them highly suitable for real SAR data processing.

(b) The principles of the WAS-GMTI mode and the DBS algorithm are researched and introduced, and a new system design scheme is presented by combining the mechanic scanning mode and the airborne SAR system, which keeps the sharpening ratio constant without increasing the complexity of either the system or the processing. To find a balance between the resolution of DBS imaging and processing efficiency, a real-time DBS algorithm based on CZT is proposed, which is able to correct the RCM and meanwhile circumvents the Doppler centroid estimation. And then, a moving target indication and parameter relocation method in mechanic scanning mode is proposed. This method indicates the targets in each DBS image, and uses the radial movements of the moving targets between the DBS images of different scans to estimate the motion parameter. The effectiveness of the proposed method is confirmed by real WAS-SAR data processing results.

(c) The nonlinearity correction and GMTI of FMCW SAR system are researched and analyzed. The principle of FMCW SAR is first introduced, and the echo model differences between the FMCW SAR and pulse SAR systems are delicately analyzed. The nonlinearity of transmitted signal in FMCW SAR is an important issue since it affects both the stationary FMCW SAR imaging quality and the moving target processing performance. The shortcoming of the classical derivative algorithm is analyzed, and then a novel nonlinearity correction algorithm based on homomorphic deconvolution is proposed, which provides accurate corrections without the restrains of the length of the controllable delay line. Based on the modeling of the moving target in FMCW SAR, a new azimuth moving target indication algorithm is proposed using the additional RWM induced by FMCW SAR system. This algorithm makes use of the additional phase, which is viewed as phase error in traditional GMTI algorithms, and therefore broadens the GMTI ability of FMCW SAR systems.

(d) Currently, the performances of most GMTIm algorithms that are effective in simulations are not satisfying in real SAR data processing. The effects of nonideal errors during the real data acquisition are analyzed, and the estimation and compensation method is provided. First of all, the motion errors of high-resolution SAR imaging are introduced, and the PGA technique is induced to correct the high-order phase error caused by these motion errors. Then, by establishing a more common moving target signal model concerning the effect of nonideal errors, the conclusion that the along-track velocity error of the aircraft and the radial velocity error can make the Doppler centroid of the moving target disperse, and deteriorate the imaging resolution. The effects of these two nonideal errors are confirmed by mathematical analysis and simulation, and a nonideal error estimation and compensation method is proposed. Together with the proposed GMTI and GMTIm algorithms in this book, a whole moving target processing scheme is presented, which is very robust and highly suitable for real SAR data processing.

Beijing, China Jian Yang

Parts of this thesis have been published in the following journal articles:

[1] Jian Yang, Chang Liu, and Yanfei Wang. Detection and imaging of ground moving targets with real SAR data. IEEE Transactions on Geoscience and Remote Sensing, vol. 53, no. 2, Feb. 2015.
[2] Jian Yang, Chang Liu, and Yanfei Wang. Imaging and parameter estimation of fast moving targets with single-antenna SAR. IEEE Geoscience and Remote Sensing Letters, vol.11, no. 2, Feb. 2014.
[3] Jian Yang, Chang Liu, and Yanfei Wang. Nonlinearity correction of FMCW SAR based on homomorphic deconvolution. IEEE Geoscience and Remote Sensing Letters, vol. 10, no. 5, Sept. 2013.
[4] Jian Yang, Chang Liu, and Yanfei Wang. Indication and imaging of along-track moving target in frequency modulated continuous wave synthetic aperture radar systems. Remote Sensing Letters, vol. 4, no. 9, pp. 891–899, Jun. 2013.
[5] Jian Yang, Chang Liu, and Yanfei Wang. Real-time DBS imaging algorithm based on chirp z-transform. IEICE Electronics Express, vol. 9, no. 21, pp. 1660–1665, Nov. 2012.
[6] Jian Yang, Chang Liu, and Yanfei Wang. Adaptive Doppler centroid estimation algorithm of airborne SAR. IEICE Electronics Express, vol. 9, no. 13, pp. 1135–1140, Jul. 2012.
[7] Jian Yang, Chang Liu, and Yanfei Wang. Ground moving target indication in single-antenna airborne SAR. The fifth Youth Academic Forum of Institute of Electronics, Chinese Acadmey of Sciences, 2013. 9.23–24.
[8] Leye Wang, Jian Yang, and Xianghui Xu. Fast moving target indication and imaging in FMCW SAR. Science technique and Engineering. 2013, vol. 13, no. 16, pp. 4534–4540.

Acknowledgement

I would like to express my utmost gratitude to my Ph.D. supervisor Prof. Yanfei Wang for his useful advices during my research. I would also like to put this book as a gift for my beloved wife Dr. Lingna Lv and our newborn son.

Contents

Abbreviations

1-D	One-Dimensional
2-D	Two-Dimensional
ATI	Along-Track Interferometry
CS	Chirp Scaling
CZT	Chirp Z-Transform
CPI	Coherent Pulse Integration
CFAR	Constant False Alarm Rate
DFT	Discrete Fourier Transform
DPCA	Displaced Phase Center Antenna
DBS	Doppler Beam Sharpening
FFT	Fast Fourier Transform
FrFT	Fractional Fourier Transform
FLOS	Fractional Low-Order Statistics
FMCW	Frequency Modulated Continuous Wave
GMTIm	Ground Moving Target Imaging
GMTI	Ground Moving Target Indication
INS	Inertial Navigation System
InSAR	Interferometric Synthetic Aperture Radar
IFFT	Inverse Fast Fourier Transform
LFM	Linear Frequency Modulated
LOS	Line-of-Sight
LUMV	Linear Unbiased Minimum Variance
MLE	Maximum-Likelihood Estimation
PGA	Phase Gradient Autofocus
PFT	Polynomial Fourier Transform
PPS	Polynomial Phase Signal
PRI	Pulse Repetition Interval
RCM	Range Cell Migration
RCMC	Range Cell Migration Correction
RD	Range Doppler

RWM	Range Walk Migration
RDM	Reflectivity Displacement Method
RVP	Residual Video Phase
SDE	Sign Doppler Estimator
SCR	Signal-to-Clutter Ratio
SNR	Signal-to-Noise Ratio
STAP	Space-Time Adaptive Processing
SAR	Synthetic Aperture Radar
UAV	Unmanned Vehicle
VCOs	Voltage-Controlled Oscillators
WAS	Wide-Area Surveillance
WVD	Wigner-Ville Distribution

Chapter 1
Introduction

Abstract Along with the development of SAR technique, imaging function cannot satisfy the requirement of the application of SAR systems in military and civil areas. Multi-function has become an important tendency of SAR. In this chapter, the development of SAR systems, GMTI and GMTIm algorithms are introduced, and the structure of this book is shown.

1.1 Development of SAR

SAR is originally designed as an advanced imaging sensor of high resolution, which is able to provide a high resolution 2-D radar image of a wide swath [1, 2]. SAR system acquires a high range resolution by transmitting and compression a large time-bandwidth product LFM signal, and achieves a high azimuth resolution by using the synthetic aperture technique [3–5]. Compared with optical remote sensing, SAR has the advantages of all-time, all-weather, and is also able to detect covered targets with the advantages of microwaves. By receiving and processing echoes with different polarizations, SAR is also capable of classifying and recognizing targets with different surface structures and materials [6, 7]. Based on these advantages, SAR is widely-used in civil and military areas [8].

SAR technique is firstly invented in the 50s, 20th century. Carl Wiley from Goodyear Company proposed the idea of increasing cross-track resolution by analyzing the Doppler spectrum of the echo on June, 1951. This idea is then named as DBS [9]. In 1953, Illinois University proved the effectiveness of DBS technique by non-coherent radar experiment. Meanwhile, another team from this university acquired the first un-focused SAR image on July, 1953 [10]. Willow Run Lab from Michigan University and the U.S. Army collaborated on the SAR system research and acquired the first focused SAR image by using the optical signal processing on August, 1957. Since then, the concept and principle of SAR was widely-known, and the application of SAR system was developed.

Based on the platform, SAR systems can be classified into spaceborne and airborne SAR. Along with the improvement of radar systems, other platforms, such

J. Yang, *Study on Ground Moving Target Indication and Imaging Technique of Airborne SAR*, Springer Theses, DOI 10.1007/978-981-10-3075-8_1

as UAV-borne SAR [11, 12], aeroboat-borne SAR [13] and missile-borne SAR [14–16], are also developed. The research on spaceborne SAR started on 70s, 20th century. The first SAR satellite Seasat-A [17, 18] was launched on May, 1978 by NASA, which was capable of providing radar images from space [19–21]. Afterwards, SIR-A, SIR-B, SIR-C/X and Lacrosse series [22] were launched by the U.S. On March 1991, Almaz-1 SAR was launched by Soviet Russia. ERS-1, ERS-2 and Envisat-1 [23, 24] were launched by the European Space Agency (ESA). JERS-1 and L-band PALSAR were launched on February 1992 and 2006 by Japan, respectively. RadarSat-1 [25, 26] and RadarSat-2 [27, 28] were launched by Canada on 1995 and 2007, respectively. In 2007, German successfully launched TerraSAR-X [29–31], and are researching on the Tandem system [32]. China, Italia, Israel, and India have also developed their own spaceborne SAR systems. The advantages of spaceborne SAR include: firstly, spaceborne SAR has wide swath, which is highly suitable for the ocean observation; secondly, the orbit of spaceborne SAR is steady, which is suitable for InSAR and Spotlight SAR modes.

Compared with spaceborne SAR, airborne SAR has the advantages of flexible working mode and simple system. Moreover, airborne SAR can be utilized to observe the interested area repeatedly without the constriction of revisit time. In 1966, a L-band airborne SAR was invented by Jet Propulsion Laboratory, which is installed on CV-900 of NASA. After sixty years' development, the performance of airborne SAR is high improved, the resolution has been rise from 15 to 1 m [33], and new working modes, such as InSAR, scan-SAR, spotlight SAR and polarize SAR have been invented. At present, many countries, such as U.S., German, Canada and China have their own airborne SAR systems, for instance, Lynx, HiSAR, AN/APG-76 of U.S., E-SAR, PAMIR of German, and so forth. Up to now, the airborne SAR is still the most widely-used SAR system in the world.

China started its research on SAR systems in 70s, 20th Century. The first SAR image was obtained by Institute of Electronics, Chinese Academy of Science in 1979. Tsinghua University, Beijing University of Aeronautics and Astronautics, Beijing Institute of Technology and Xidian University have also published lots of researches on SAR technique. With these researches, the performances of SAR systems in China are improved, the finest resolution is up to 0.1 m [34].

Along with the development of the hardware and signal processing technology, the developing tendencies of SAR systems are as follows.

(a) High Resolution

High resolution is a lasting pursuit of the SAR system. Typical high-resolution SAR system includes Lynx [35, 36], MiniSAR [37], RAMESE [38], PAMIR [39–42] and so forth. The finest resolution of an airborne SAR in China is 0.1 m, which is achieved by band-composing. At present, the high resolution SAR systems are mainly airborne SAR systems. Spaceborne SAR systems have relatively lower resolutions than airborne SAR systems, which are higher than 1 m.

Along with the improvement of the image resolution, the detection, recognition, classification and description of the targets can be achieved. Therefore, high resolution SAR images can be used not only on mapping domain but also on moving target indication and recognition [43].

(b) Multifunction

SAR system is originally designed as an imaging sensor. However, with the increasing requirements, SAR systems with multiple functions and working modes are becoming the research hotspot. The working mode includes stripmap SAR, scan-SAR, spotlight SAR [44–46], sliding spotlight SAR [47–49] and so forth. Different working modes can satisfy different requirements of resolutions and observation swath. Besides imaging, other functions such as GMTI and InSAR [50–52] are required. Phased-array antenna technique is also used in SAR system, which makes the multifunction and fast switch possible [53–55].

(c) Miniaturization

Along with the development of micro-electronics and antenna techniques, small volume, light weight, and low power consumption SAR systems are becoming a tendency of the SAR system. Mini SAR system can be applied on UAV, which is a burgeoning area in military applications. Typical systems include MiSAR [56, 57], MiniSAR, μSAR [58, 59] with FMCW technique and so forth.

(d) Multi-band, Multi-polar

Microwaves of different bands have different transmission capabilities. Therefore, more information can be derived by the fusion of images of different bands. Material and surface structure information can be extracted from the polarized SAR images, and is used in target recognition and classification [60]. Therefore, multi-band, multi-polar SAR is an important tendency of the SAR technique.

1.2 Development of GMTI and GMTIm Techniques

1.2.1 Technique Backgrounds

The concept of radar can be traced back to the famous microwave experiment of German physicist Hertz at the end of 19th century. The principle of radar is to detect and locate the target by using the scattering of microwave. Radar was first applied in the World War II in the airplane and warship surveillance and detection. The range of a target is acquired by the time-delay of the received signal, and the direction of a target is acquired by the scan of the antenna [61–63]. In conclusion, the detection of moving targets are the origin function of radar.

As an imaging radar, SAR was firstly designed with only the function of imaging. The signal used in the stationary scene imaging of SAR is called the clutter in the traditional radar. However, the combination of radar imaging and moving target processing can acquire more information of the observation area.

With the advantage of all-time and all-weather, SAR system can provide both the high resolution radar image of the observation area and the location of small moving target such as military vehicles, mobile missile launcher, and artillery.

GMTI is an important function of modern radar system. This function was firstly applied in AN/APG-76, which is capable of real-time imaging of stationary scene and indicating moving targets [64, 65]. The U.S. land forces and air forces set up the Joint STARS, which played an important role in the Iraq War [66]. Lynx system is boarded on I-GNAT, which can provide high resolution images and indicate moving targets with the slowest velocity of 2.98 m/s within the scope of 270°. HiSAR system is boarded on Global Hawk, which has both WAS mode and stripmap SAR/GMTI mode. The HiSAR system has a working range of 20–110 km, observation swath of 37 km, and has a resolution of 6 m in the stripmap SAR/GMTI mode [67]. Other typical airborne SAR systems with GMTI functions include MiniSAR, TESAR on Predator, EL/M2055, EL/M2060P, EL/M2040T, PAMIR, ASTOR. GMTI is also used on spaceborne SAR. The most famous SAR satellites are Lacrosse series of the U.S. and the RadarSat-2 of Canada. However, spaceborne SAR system cannot detect slow moving targets, which makes it not suitable for GMTI.

GMTIm is to obtain a 2-D high resolution of the moving target after it is detected. With the image of the target, the moving target can be further recognized and classified. However, in the existence of unknown motions, moving targets are smeared and dis-located in the stationary SAR image. GMTIm algorithms have to accurately estimate the motion parameters and eliminate the interference of system errors [68]. Many researchers have devoted to research the GMTIm algorithms, but few have satisfying performances in real data processing.

1.2.2 Recent Development of GMTI and GMTIm Technique

GMTI and GMTIm are important functions of SAR system. However, since the moving targets have additional motions, two problems must be solved in the SAR moving target processing.

First, the speeds of airborne and spaceborne SAR are fast, which lead to the Doppler bandwidth broaden of the clutter. Targets with slow cross-track velocities will be submerged by the clutter, and is hard to be detected [69]. Therefore, the suppression of the energy of the clutter must be done in GMTI.

Second, the imaging principle of SAR is to use the CPI between the target and the scene. To the stationary scene, the relative motions are decided by the radar platform, and the stationary SAR image can be obtained after the phase error is compensated [70]. However, as to the moving target, the relative motions are composed by both the motions of the platform and the target. Since the motion of the target is unknown, stationary SAR imaging algorithms cannot fully compensate the phase error, which will lead to the smear and dis-locate of the target, as shown in Fig. 1.1 [35, 36].

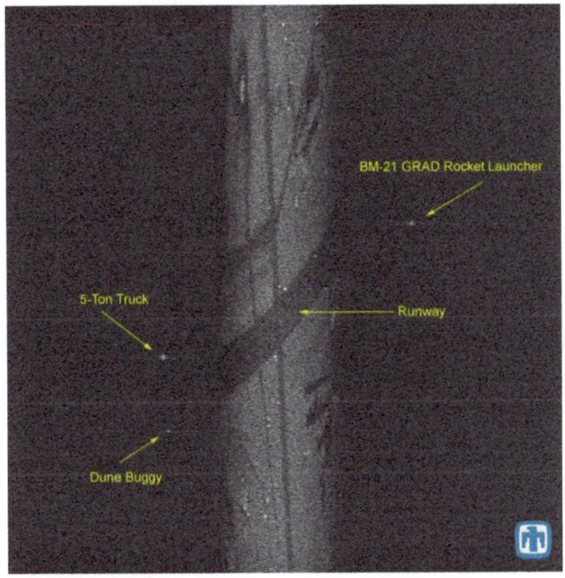

Fig. 1.1 Image of moving targets of Lynx system

The main processing steps of GMTI and GMTIm of SAR are as follows.

(a) Suppress the clutter, increase the SCR, and set judge strategy detect the moving targets;
(b) Extract the echo of the moving target, and estimate the Doppler and motion parameters;
(c) Refocus the moving target, and locate the image of the target in its actual location in the image.

The differences between the Doppler parameters of the clutter and the moving target are not only the cause of the defocus and the dis-location, but also the lead to indicate and focus the target. In 40 years, researchers have proposed many GMTI and GMTIm algorithms. These algorithms can be classified into single-channel and multi-channel algorithms. Single-channel algorithms include

(a) Spectrum filtering algorithm [71–73]. The cross-track velocity of a moving target will induce a Doppler centroid shift to the echo. Based on the Doppler centroid shift, the moving targets whose Doppler centroid locates out of the spectrum of the clutter can be detected, and the motion parameters can be estimated from the shift.
(b) Shear average algorithm [74] and RDM algorithms [75]. The along-track velocity of a moving target will induce a change of Doppler modulation rate. The Doppler modulation rate of an along-track moving target is different from that of the clutter. These two algorithms use this feature to detect moving targets and estimate the motion parameters.

(c) Time-frequency analysis algorithm: The Doppler characters of the moving target echo are used to detect the target by time-frequency analysis algorithms, such as Wigner-Ville Distribution (WVD) algorithm [76–79].
(d) Sub-aperture algorithm [80]: This sort of algorithms uses the different locations of the moving target in the stationary sub-aperture SAR images to detect targets and estimate motion parameters.

Other single-channel GMTI algorithms include Parameter searching algorithm [81], Keystone transform algorithm, Wavelet transform algorithm [82, 83] and so forth. All these algorithms are based on the different Doppler characters of the moving target. However, constrained by the single-antenna system, the single-antenna has a weaker ability in slow moving target indication. Therefore, along with the hardware development and signal processing ability, multi-channel GMTI algorithms are developed.

(a) DPCA [84–86]: Multiple phase centers are placed along the azimuth direction by DPCA, and by elimination in the complex image domain, the moving target can be extracted, and the motion parameters can be estimated. This algorithm has a high requirement of the location of multiple receivers, so it is limited in practical applications.
(b) ATI [87–89]: Same as DPCA algorithm, ATI also needs multiple phase centers that settled along the azimuth direction. However, the ATI algorithm uses the interference of two isolated echoes to eliminate the clutter, and is becoming the main multi-channel GMTI algorithm in practical applications.
(c) STAP [90–92]: This algorithm uses the multiple sets of signals obtained by antenna-array, and suppresses the clutter by joint processing of space-time signal. Theoretically, the STAP algorithm has the best performance in GMTI. However, the heavy calculation burden limits its applications.

Besides the traditional stripmap SAR mode, another widely-used mode, i.e., the wide area surveillance (WAS) mode is applied in SAR [93]. By scanning the antenna along the azimuth direction, WAS mode is able to obtain a larger observation area, and the revisit of moving target can be achieved.

FMCW SAR is another new working scheme of SAR. By combining the FMCW technique and SAR system, FMCW SAR has the advantages of small size, light weight, low cost and high resolution [94]. FMCW SAR is highly suitable for small platforms, and has a promising performance in GMTI. With the development in FMCW SAR technique, the FMCW SAR has become a hotspot in SAR research area.

1.3 Book Structure and Organization

Up to now, single-antenna airborne SAR is still the most widely-used SAR system in the world. The single-antenna airborne SAR GMTI and GMTIm algorithms are the main topics in this book. Three working modes, including stripmap SAR, WAS mode and FMCW SAR, are studied, and the non-ideal motion error in practical SAR signal processing is analyzed. This book is organized in six chapters, and the main content of each chapter is as follows.

This chapter is the introduction, the development and principle of SAR systems, GMTI algorithms and GMTIm algorithms are introduced in this chapter, and the book structure is given.

Chapter 2 contains two main parts. First, the signal model of a moving target in airborne SAR is established, and impacts of the motions on the echo character are analyzed. Second, the principles of traditional GMTI and GMTIm algorithms are introduced, and the problems and restrains of these algorithms are analyzed.

Chapter 3 studies the GMTI and GMTI of fast moving targets in stripmap SAR. Firstly, the signal model of fast moving targets in stripmap SAR/GMTI mode is established, and the impacts of fast cross-track velocities on the Doppler ambiguity and higher order azimuth phase are analyzed. Then, A classification of targets by the location of their spectra is presented, based on which a Doppler centroid estimation algorithms based on curve fitting, a multiple target indication and echo extraction method, and a fast moving target imaging algorithm based on Hough transform and third-order PFT are presented. Finally, simulations and real data are utilized to prove the effectiveness of these algorithms.

Chapter 4 studies the GMTI algorithms in WAS mode. The signal mode of moving target is established in WAS mode, and a new system design scheme is presented to maintain the sharpening ratio constant without increasing the system and signal processing complexity. Moreover, a real-time DBS algorithm is proposed based on CZT, and a GMTI and motion parameter estimation algorithm based on multiple revisits is presented. Finally, simulations and real data are utilized to prove the effectiveness of these algorithms.

Chapter 5 studies the nonlinearity correction and GMTI algorithms in FMCW SAR. The differences between a FMCW SAR and a pulse SAR are analyzed, and additional phase error of FMCW SAR is calculated. Furthermore, a nonlinearity correction algorithm based on Homomorphic Deconvolution is proposed, and an along-track moving target indication algorithm is presented. By using the additional phase error, the GMTI performance of FMCW SAR is improved.

Chapter 6 researches the non-ideal motion errors in practical SAR signal processing. Pointing at the problem that existing GMTIm algorithms have poor performances in the real SAR data processing, the impact of Doppler centroid is analyzed. The signal model of moving targets with non-ideal motion error is established, and the platform velocity error and cross-track velocity error of the moving target are analyzed. An error estimation and compensation algorithm is presented, and a whole practical SAR data processing scheme with the algorithms in the frontal chapters is proposed. Finally, simulations and real data are utilized to prove the effectiveness of these algorithms.

References

1. Liu Y (2003) Radar imaging technology. Harbin Institute of Technology Press, Harbin
2. Zhang Z (1989) Principle, system and application of synthetic aperture radar. Science Press, Beijing
3. Bao Z, Xing M, Wang T (2005) Radar imaging technology. Publishing House of Electronics industry, Beijing
4. Mehrdad S (1999) Synthetic aperture radar signal processing with Matlab algorithms. Wiley
5. Curlander JC, McDonough RN (2006) Synthecia aperture radar–systems & signal processing. Publishing House of Electronics industry, Beijing
6. Qu C, He Y, Gong S (2002) Development of airborne SAR. Mod Radar 24(1):1–14
7. William MB (1967) Synthetic aperture radar. IEEE Trans Aerosp Electron Syst 3(2):217–229
8. Wiley CA (1985) Synthetic aperture radar. IEEE Trans Aerosp Electron Syst 21(3):440–443
9. Sherwin CW, Ruina JP, Raweliffe RD (1962) Some early developments in synthetic aperture radar system. IRE Trans Mil Electron 6(2):111–115
10. Skolnik MI (1985) Fifty years of radar. Proc IEEE 73:182–197
11. Wang J, Liu J, Yuan Y et al (2009) Research on application of unmanned aerial vehicles borne SAR. In: APSAR, 2009, pp 60–63
12. Li Y, Liang F, Song Q et al (2011) Lever arm rotation compensation for UAV mounted SAR. In: APSAR, 2011, pp 1–3
13. Wang Y, Xu X, Liu C et al (2010) Airship SAR system for precision microwave measurement of ground targets. J Electron Inf Technol 32(1):28–31
14. Zhou Q, Qu C, Su F (2008) A new approach of extended Chirp Scaling algorithm for high squint missile-borne SAR data processing. ISCSCT 2:133–136
15. Yin C, Jia X, Qin W (2009) A novel algorithm for missile borne linear array antenna synthetic aperture radar imaging. In: ICIEA, 2009, pp 1488–1492
16. Yi Y, Zhang L, Li Y et al (2009) Range Doppler algorithm for bistatic missile-borne forward-looking SAR. In: APSAR, 2009, pp 960–963
17. Jordan R (1980) The Seasat-A synthetic aperture radar system. IEEE J Oceanic Eng 5(2): 154–164
18. Thompson T, Laderman A (1976) Seasat-A synthetic aperture radar. Radar System Implementation. OCEANS 8:247–251
19. Zhao Y (2007) Ground moving target indication of synthetic aperture radar. Institute of Electronics, Chinese Academy of Sciences
20. Li Y (2012) Motion error estimation and compensation of airborne SAR. Institute of Electronics, Chinese Academy of Sciences
21. Hu J (2012) Signal processing and imaging in multi-channel SAR systems. Institute of Electronics, Chinese Academy of Sciences
22. Wehner DR (1995) High-resolution radar, 2nd edn. Artech House, Norwood
23. Mancini P, Suchail JL, Desnos YL et al (1996) The development of the Envisat-1 advanced synthetic aperture radar. In: IEEE international geoscience and remote sensing symposium, 1996, pp 1355–1357
24. Desnos YL, Laur H (1999) The Envisat-1 advanced synthetic aperture radar processor and data products. In: IEEE international geoscience and remote sensing symposium, 1999, pp 1683–1685
25. Julien C, Monique B, Gaetan L (2003) RADARSET-1 SAR scenes for wind power mapping in coastal area: Gulf of St-Lawrence case. In: IEEE international geoscience and remote sensing symposium, 2003, pp 2700–2702
26. Steven I, Greta B (2004) RADARSAT-1: Canadian space agency hurricane watch program. In: IEEE international geoscience and remote sensing symposium, 2004, pp 2742–2745
27. Livingstone C (1998). The addition of MTI modes to commercial SAR satellites. In: Proceedings of 10th CASI conference on Astronautics, Ottawa, Canada, 1998, pp 26–28
28. Luscombe A (1995) The Radarsat project. In: IEEE Canadian Review, 1995

29. Buckreuss S, Balzer W, Muhlbauer P (2003) The TerraSAR-X satellite project. In: IEEE international geoscience and remote sensing symposium, 2003, vol 5, pp 3096–3098
30. Stangl M, Werninghaus R, Schwerizer B et al (2006) TerraSAR-X technologies and first results. IEE Proc Radar Sonar Navig 153(2):86–95
31. Marco S, David H, Benjamin B et al (2005) TerraSAR-X: Calibration concept of a multiple mode high resolution SAR. In: IEEE international geoscience and remote sensing symposium, 2005, pp 4874–4877
32. Alberto M, Gerhard K, Irena H et al (2004) TanDEM-X: A TerraSAR-X add-on satellite for single-pass SAR interferometry. In: IEEE international geoscience and remote sensing symposium, 2004, pp 1000–1003
33. Zhang M (2012) Motion error compensation of multi-channel SAR system. Institute of Electronics, Chinese Academy of Sciences
34. Wu X (2012) Signal processing and imaging of random noise radar. Institute of Electronics, Chinese Academy of Sciences
35. "Lynx Synthetic Aperture Radar", http://www.sandia.gov/radar/lynx.html
36. Tsunoda SI, Pace F, Stence J, Woodring M (2000) Lynx: A high-resolution synthetic aperture radar, pp 51–58
37. Sweet AD, Dubbert DF, Doerry AW, Sloan GR, Dee Gutierrez GR (2006) A portfolio of fine resolution Ku-band miniSAR images: part I. In: Proceedings of SPIE 6210. Radar Sensor Technology X, 6210, May 2006
38. Cantalloube HMJ, Fernandez PD (2003) Airborne X-band SAR imaging with 10 cm resolution—technical challenge and preliminary results. In: IEEE international geoscience and remote sensing symposium, 2003, pp 185–187
39. Brenner AR, Roessing L (2008) Radar imaging of urban areas by means of very high-resolution SAR and interferometric SAR. IEEE Trans Geosci Remote Sens 46 (10):2917–2982
40. Ender JHG, Brenner AR (2003) PAMIR-A wideband phased array SAR/MTI system. IEE Proc Radar Sonar Navig 150(3):165–172
41. Cerutti-Maori D, Klare J, Brenner AR et al (2008) Wide-area traffic monitoring with the SAR/GMTI system PAMIR. IEEE Trans Geosci Remote Sens 46(10):3019–3030
42. Brenner AR, Ender JHG (2006) Demonstration of advanced reconnaissance techniques with the airborne SAR/GMTI sensor PAMIR. IEEE Trans Geosci Remote Sens 153(2):152–162
43. Brenner AR (2010) Proof of concept for airborne SAR imaging with 5 cm resolution in X-band. In: EUSAR, 2010, pp 615–618
44. Jakowatz CV, Wahl DE, Eichel PH et al (1996) Spotlight-mode synthetic aperture radar: a signal processing approach. Kluwer Academic Publishers, Boston
45. Carrara WG, Goodman RS, Majewdki RM (1995) Spotlight synthetic aperture radar-signal processing algorithms. Artech House, Norwood
46. Zhu D, Zhu Z (2007) Range Resampling in the polar format algorithm for spotlight SAR image formation using the Chirp-Z transform. IEEE Trans Signal Process 55(3):1011–1023
47. Belcher DP, Baker CJ (1995) High resolution processing of hybrid strip-map/spotlight mode SAR. IEE Proc Radar Sonar Navig 143(6):366–374
48. Franceschetti G, Guida R, Iodice A et al (2004) Efficient simulation of hybrid stripmap/spotlight SAR raw signals from extended scenes. IEEE Trans Geosci Remote Sens 42(11):2385–2396
49. Lanari R, Zoffoli S, Sansosti E et al (2001) New approach for hybrid strip-map/spotlight SAR data focusing. IEE Proc Radar Sonar Navig 148(6):363–372
50. Ferraiuolo G, Meglio F, Pascazio V et al (2009) DEM reconstruction accuracy in multichannel SAR interferometry. IEEE Trans Geosci Remote Sens 42(11):2385–2396
51. Ferretti A, Prati C, Rocca F (1999) Multibaseline InSAR DEM reconstruction: the wavelet approach. IEEE Trans Geosci Remote Sens 37(2):705–715
52. Lanari R, Fornaro G, Riccio D et al (1996) Generation of digital elevation models by using SIR-C/X-SAR multifrequency two-pass interferometry: the Etna case study. IEEE Trans Geosci Remote Sens 34(5):1097–1114

53. Hansen RC (2009) Phased array antennas. John Wiley & Sons Inc, NJ
54. Li G, Xu J, Peng Y et al (2007) Location and imaging of moving targets using nonuniform linear antenna array SAR. IEEE Trans Aerosp Electron Syst 43(3):1214–1220
55. Ender JHG, Klare J (2009) System architectures and algorithms for radar imaging by MIMO-SAR. In: IEEE Radar Conference, Pasadena, CA, May 2009, pp 1–6
56. Edrich M (2004) Design overview and flight test results of the miniaturized SAR sensor MISAR. In: EURAD, Amsterdam, Netherlands, Oct 2004, pp 205–208
57. Edrich M (2006) Ultra-lightweight synthetic aperture radar based on a 35 GHz FMCW sensor concept and online raw data transmission. IEE Proc Radar Sonar Navig 153(2):129–134
58. Zaugg EC, Hudson DL, Long DG (2006) The BYU mu SAR: a small, student-built SAR for UAV operation. In: IEEE international geoscience and remote sensing symposium, 2006, pp 411–414
59. Zaugg EC, Long DG (2008) Theory and application of motion compensation for LFM-CW SAR. IEEE Trans Geosci Remote Sens 46(10):2990–2998
60. Garester F, Dubois-Fernandez PC, Guyon D et al (2009) Forest biophysical parameter estimation using L- and P-band polarimetric SAR data. IEEE Trans Geosci Remote Sens 47 (10):3379–3388
61. Ding L, Geng F (2000) Radar principle. Xidian University Press, Xi'an
62. Clarke J (1985) Airborne early warning radar. Proc IEEE 73(2):312–324
63. Morchin WC, Johnston SL (1991) Modern airborne early warning radars. Microwave J 30–47. Academic OneFile. Accessed 1 Jan 2017
64. Tobin ME, Greenspan M (1996) Adaptation of AN/APG-76 multimode radar to the smuggling interdiction mission. In: IEEE national radar conference, Ann Arbor, Michigan, May 1996, pp 13–18
65. Tobin ME, Greenspan M (1996) Smuggling interdiction using an adaptation of the AN/APG-76 multimode radar. IEEE Aerosp Electron Syst Mag 11(11):19–24
66. Richard JD (2004) Ground moving target indicator radar and the transformation of U.S. warfighting, Northrop Grumman, Feb 2004
67. Bayma RW (1996) Hughes integrated synthetic aperture radar. In: IEEE international geoscience and remote sensing symposium, 1996, vol 3, pp 1615–1617
68. Liu C (2006) Real-time signal processing of airborne SAR/GMTI systems. Institute of Electronics, Chinese Academy of Sciences
69. Zheng M (2003) Moving target indication and imaging of synthetic aperture radar. Institute of Electronics, Chinese Academy of Sciences
70. Kang X (2004) Gournd moving target indication and imaging of airborne SAR. Institute of Electronics, Chinese Academy of Sciences
71. Raney RK (1971) Synthetic aperture imaging radar and moving target. IEEE Trans Aerosp Electron Syst 7:499–505
72. Chen H, McGillem CD (1992) Target motion compensation by spectrum shifting in synthetic aperture radar. IEEE Trans Aerosp Electron Syst 28(3):119–125
73. Freeman A (1984) Simple MTI using synthetic aperture radar. In: IEEE international geoscience and remote sensing symposium, 1984, pp 65–70
74. Fienup JR (2001) Detection of moving targets in SAR imagery by focusing. IEEE Trans Aerosp Electron Syst 37(3):794–809
75. Moreira J, Keydel W (1995) A new MTI-SAR approach using the reflectivity displacement method. IEEE Trans Geosci Remote Sens 33(5):1238–1244
76. Barbarossa S (1992) Detection and imaging of moving objects with synthetic aperture radar, part I: optimal detection and parameter estimation theory. IEE Proceeds-F 139(1):79–88
77. Barbarossa S, Farina A (1992) Detection and imaging of moving objects with synthetic aperture radar, part II: joint time-frequency analysis by Wigner-Ville distribution. IEE Proceeds-F 139(1):89–97
78. Liu F, Sun D, Huang Y et al (2009) Fast parameter-estimation of LFM signal based on improved combined WVD and randomized Hough transform. ACTA Armamentarii 30 (12):1642–1646

79. Liu J, Wang X, Liu Z et al (2007) Detection performance of linear frequency modulated signals based on Wigner-Hough transform. ACTA Electronica Sin 35(6):1212–1217
80. Kirscht M (1996) Detection and velocity estimation of moving objects in a sequence of single-look SAR images. In: IEEE international geoscience and remote sensing symposium, 1996, Lincoln, Nebraska, USA, pp 333–335
81. Dias J, Marques P (2003) Multiple moving target detection and trajectory estimation using a single SAR sensor. IEEE Trans Aerosp Electron Syst 39(2):604–624
82. Li L, Si X, Chai J et al (2009) Parameters estimation for LFM radar signal based on reassigned wavelet-Radon transform. Syst Eng Electron 31(1):74–77
83. Li G, Zhu H (2001) Airborne SAR moving target parameter estimation based on extended wavelet transform. J Electron Inf Technol 23(11):1154–1161
84. Fante RL (1989) Analysis of the Displaced-Phase-Center radar for clutter reduction. Mitre Corporation Report, Aug 1989, MT10666
85. Lightstone L (1944) Antenna distortions in multiple phase centre interferometric systems. In: IEEE international geoscience and remote sensing symposium, Aug 1944, vol 4, pp 1980–1982
86. Wang HSC (1991) Mainlobe clutter cancellation by DPCA for space-based radars. In: Aerospace applications conference, 1991. Digest., 1991 IEEE, Feb 1991, pp 1–128
87. Stockburger E, Held (DN) Interferometric ground moving target imaging. In: IEEE international radar conference, 1995, 438–443
88. Soumekh M (1997) Moving target detection in Foliage using along track monopulse synthetic aperture radar imaging. IEEE Trans Image Process 6(8):1148–1163
89. Moccia A, Rufino G (2001) Spaceborne along-track SAR interferometry: performance analysis and mission scenarios. IEEE Trans Aerosp Electron Syst 37(1):199–213
90. Brennan LE, Reed IS (1973) Theory of adaptive radar. IEEE Trans Aerosp Electron Syst 9 (3):237–252
91. Ender JHG (1999) Space-time processing for multichannel synthetic aperture radar. Electron Commun Eng J 11(1):29–38
92. Gerlach K, Picciolo ML (2003) Airborne/Spacebased radar STAP using a structured covariance matrix. IEEE Trans Aerosp Electron Syst 39(1):269–281
93. Zhao F (2013) Muli-channel SAR imaging and moving target indication technique. Institute of electronics, Chinese Academy of Sciences
94. Yu B (2013) FMCW-SAR signal processing and imaging technique. Institute of electronics, Chinese Academy of Sciences

Chapter 2
SAR Ground Moving Target Indication and Imaging Theory

Abstract This chapter mainly introduces the basic principles of SAR imaging, signal model, GMTI and GMTIm algorithms. The geometries of stationary and moving targets are established, and the characters of their echoes are analyzed. Based on the analysis, three classical single-channel GMTI algorithms are introduced in this chapter, and their limitations are discussed. Moreover, the principle of GMTIm algorithms with Keystone transform and time-frequency analysis algorithm is introduced, and the limitations in Doppler centroid ambiguity and motion parameter derivation are studied.

2.1 Introduction

Signal modeling and phase analyzing are the basis of SAR GMTI and GMTIm theories. Both the characters and processing problems of moving targets in real data can be explained by the signal model. Therefore, it is necessary to study on the signal model in the beginning of this book. After modeling the echo signal of a moving target, a detailed analysis of the impact of motion parameters on the echo phase in this chapter, and the different characters of moving targets with different motions, which are both the problems and clues of GMTI and GMTIm algorithms, are discussed.

In real applications, single-channel SAR systems still outnumber multi-channel SAR systems in China, due to the high expense and complex system structure of a multi-channel SAR system. In the signal processing point of view, the performances of multi-channel GMTI algorithms are higher due to the expense of complex system structures and algorithms. Although single-channel GMTI algorithms have been studied for years, there are many problems in real applications, considering the unknown and complex motions of moving targets. Compared with spaceborne SAR, airborne SAR has a lower speed and a higher resolution, which endows airborne SAR a lower detectable speed and a finer indication pixel. Therefore, this book mainly focuses on the single-channel GMTI and GMTIm algorithms of airborne SAR systems.

J. Yang, *Study on Ground Moving Target Indication and Imaging Technique of Airborne SAR*, Springer Theses, DOI 10.1007/978-981-10-3075-8_2

In this chapter, the signal model of a moving target is established, and the principles of several traditional single-channel GMTI and GMTIm algorithms are introduced, which is organized as follows. In Sect. 2.2, the principle of a SAR system is briefly introduced, and the signal model of a stationary target is established. In Sect. 2.3, the signal model of a moving target is established, and the relationship between the echo property and the motions is deeply analyzed. In Sects. 2.4 and 2.5, traditional single-channel SAR GMTI and GMTIm algorithms are presented respectively, and their limitations are discussed. Finally, conclusive remarks are provided in Sect. 2.6.

2.2 Principle of SAR

SAR is an advanced 2-D high resolution radar, which is able to acquire high imaging resolutions in both the range and azimuth directions. For a real aperture radar, the azimuth resolution relies on both the size of the aperture and the range: bigger aperture and shorter range make higher azimuth resolution. Therefore, in order to improve the azimuth resolution, the aperture size must be enlarged. The innovation of SAR is to simulate an equivalent large azimuth aperture by moving the radar in the azimuth direction with a constant velocity. On the other hand, the high range resolution of SAR is obtained by transmitting and receiving a wide-band linear modulated (LFM) signal.

During the azimuth time t_a, the platform flies from O to O_1 with a constant velocity V_a. R_0 and $R(t_a)$ represent the nearest and instantaneous slant ranges between the platform and the target P, respectively. H and X represent the height of the platform and the flat range, respectively. According to Fig. 2.1, $R(t_a)$ can be expressed as

$$R(t_a) = \sqrt{R_0^2 + (V_a t_a)^2} \tag{2.2.1}$$

Expand Eq. 2.2.1 into a Taylor series and keep it to the second-order term of t_a, then Eq. 2.2.1 can be approximated as

$$R(t_a) \approx R_0 + \frac{V_a^2}{2R_0} t_a^2 \tag{2.2.2}$$

Suppose the transmitted signal is a LFM signal with a carrier frequency f_0, a modulation rate K_r and a pulse band-width T_r, then the transmitted signal $s_t(t_r)$ can be expressed as

$$s_t(t_r) = rect\left(\frac{t_r}{T_r}\right) \exp\left[j2\pi\left(f_0 t_r + \frac{K_r}{2} t_r^2\right)\right] \tag{2.2.3}$$

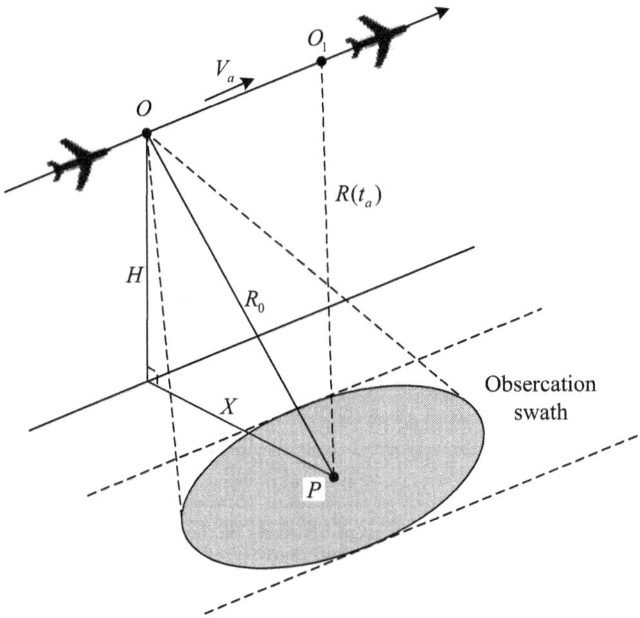

Fig. 2.1 Broadside geometry of a stationary target in airborne SAR

where $rect(\cdot)$ represents the rectangle window of the transmitted signal, and t_r denotes the range time. After the range compression, the echo signal of target P can be expressed as

$$s_r(t_a,t_r) = A_0\omega_a(t_a)\sin c\left[\pi B_r\left(t_r - \frac{2R(t_a)}{c}\right)\right]\exp\left[-j\frac{4\pi}{\lambda}R(t_a)\right] \tag{2.2.4}$$

where A_0 is the complex reflection coefficient of the moving target, B_r and λ are the bandwidth and wavelength of the transmitted signal, respectively, and c is the speed of light.

From Eq. 2.2.4, the range phase information is contained in the sinc function, which indicates that the range phase is related to the instantaneous slant range $R(t_a)$. The azimuth phase information is contained in the exponential terms, which means that it is also related to $R(t_a)$. Substitute Eq. 2.2.2 into Eq. 2.2.4, it yields

$$\begin{aligned} s_r(t_a,t_r) = & A_0\omega_a(t_a)\sin c\left[\pi B_r\left(t_r - \frac{2R_0}{c}\right)\right]\exp\left(-j\frac{4\pi}{\lambda}R_0\right)\exp\left(-j\frac{4\pi}{\lambda}\frac{V_a^2}{2R_0}t_a^2\right) \\ & \cdot\exp\left(-j\frac{4\pi K_r}{c}\frac{V_a^2}{2R_0}t_r t_a^2\right) \end{aligned} \tag{2.2.5}$$

In Eq. 2.2.5, the first exponential term denotes the range location of the target; the second exponential term represents the azimuth phase of the target. It can be noted that the azimuth phase of a point target in stripmap SAR is a second-order function of t_a, which makes it a LFM signal with an azimuth modulation rate f_{dr} as

$$f_{dr} = \frac{2V_a^2}{\lambda R_0} \tag{2.2.6}$$

Therefore, the azimuth compression can be achieved by utilizing a matched filter with modulation rate f_{dr}. If the modulation rate of the matched filter is mismatched, the second-order azimuth phase cannot be accurately compensated, and the azimuth dispersion is induced.

The third exponential term of Eq. 2.2.5 represents the RCM of the target. The RCM contains the range curve migration and the RWM. In broadside stripmap SAR, there is only the range curve migration with the absence of the squint angle. The RCM error will cause an energy dispersion of a target among multiple range cells if not accurately compensated.

According to the analysis above, the core of SAR signal processing is to compensate the phase errors of both the range and azimuth directions. The range and azimuth phase errors are compensated by using the matched filters. Since the transmitted signal is known, the range matched filter is mostly accurate. The azimuth matched filter is calculated and estimated, so it may induce azimuth phase errors if the matched filter is not accurate. The first-order azimuth phase error will cause the dislocation of the target; the second-order azimuth phase error will cause the broadening of the main-lobe; the third-order azimuth phase error will cause the asymmetry of the side-lobes; the fourth-order azimuth phase error will cause the raise of the side-lobes. RCMC is also an important part. Different algorithms have different ways to accomplish the RCMC.

As to a moving target, the signal form is the same as that of a stationary target, while the phase information is different with the existence of the motion. Therefore, the imaging algorithms that suit the stationary target processing are also suitable for the moving targets, if only the imaging parameters are accurately estimated.

2.3 Signal Analysis of Moving Target in SAR

2.3.1 Signal Model of a Moving Target

In a typical airborne SAR system, the synthetic aperture time is several seconds. During the synthetic aperture time, a target is radiated and reflected the microwave continuously. In most articles, the moving target is assumed to be moving with a constant velocity or a constant acceleration during the synthetic aperture time. According to this assumption, A broadside SAR configuration with a moving target in the slant range domain is shown in Fig. 2.2.

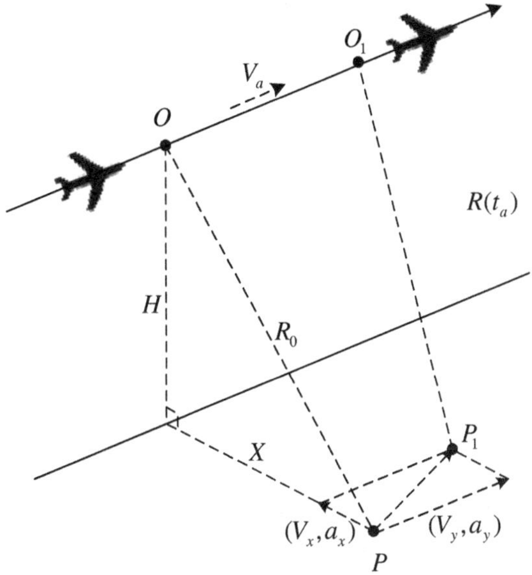

Fig. 2.2 Broadside geometry of a moving target in a single-antenna SAR system

The platform parameters are the same as in Fig. 2.1, whereas the target is a moving target with cross-track velocity V_x, cross-track acceleration a_x, along-track velocity V_y and along-track acceleration a_y. During the azimuth time t_a, the platform flies from O to O_1 with a constant velocity V_a, while the moving target moves from P to P_1. Therefore, the instantaneous slant range $R(t_a)$ can be expressed as

$$R(t_a) = \sqrt{H^2 + \left(X - V_x t_a - \frac{1}{2} a_x t_a^2\right)^2 + \left(V_a t_a - V_y t_a - \frac{1}{2} a_y t_a^2\right)^2} \tag{2.3.1}$$

Expand Eq. 2.3.1 into a Taylor series and keep it to the second-order term of t_a, then Eq. 2.3.1 can be approximated as

$$R(t_a) \approx R_0 - \frac{X}{R_0} V_x t_a + \frac{(V_a - V_y)^2 - a_x X}{2R_0} t_a^2 \tag{2.3.2}$$

The relationship of the motion parameters between the slant range domain and range domain can be expressed as

$$\frac{X}{R_0} = \frac{V_r}{V_x} = \frac{a_r}{a_x} \tag{2.3.3}$$

where V_r and a_r represent cross-track velocity and acceleration in the slant-range domain respectively as shown in Fig. 2.3. Substitute Eq. 2.3.3 into Eq. 2.3.2, it yields

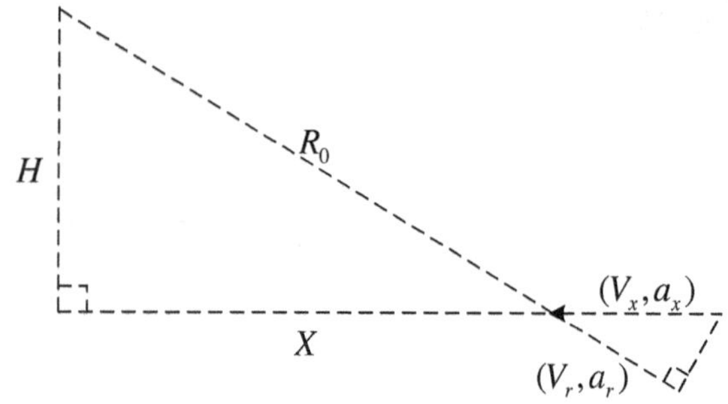

Fig. 2.3 Projection relationship between the range and slant-range domain

$$R(t_a) \approx R_0 - V_r t_a + \frac{(V_a - V_y)^2 - a_r R_0}{2R_0} t_a^2 \tag{2.3.4}$$

Substitute Eq. 2.3.4 into Eq. 2.2.4, the echo signal of the moving target can be expressed as

$$\begin{aligned} s_r(t_a, t_r) = {} & A_0 \omega_a(t_a) \sin c\left[\pi B_r\left(t_r - \frac{2R_0}{c}\right)\right] \exp\left(\mathrm{j}\frac{4\pi}{\lambda} V_r t_a\right) \exp\left(-\mathrm{j}\frac{4\pi}{\lambda} \frac{(V_a - V_y)^2 - R_0 a_r}{2R_0} t_a^2\right) \\ & \cdot \exp\left(\mathrm{j}\frac{4\pi K_r}{c} V_r t_r t_a\right) \exp\left(-\mathrm{j}\frac{4\pi K_r}{c} \frac{(V_a - V_y)^2 - R_0 a_r}{2R_0} t_r t_a^2\right) \exp\left(-\mathrm{j}\frac{4\pi}{\lambda} R_0\right) \end{aligned} \tag{2.3.5}$$

Compare Eq. 2.3.5 with Eq. 2.2.5, it is noted that the signal model of the moving target is different from that of a stationary target. The first exponential term of Eq. 2.3.5 is the Doppler centroid phase, which results in the azimuth dis-location of the moving target. The second exponential term is the second-order azimuth phase, it is also different from a stationary target, indicating that the Doppler modulation rate of a moving target differs from that of a stationary target. The third and fourth exponential terms are the RCM of the moving target. Compared with a stationary target, an additional RWM is induced by the motions, and the range curve migration is different from a stationary target. The RCM will deteriorate the imaging resolution if not accurately compensated, which is also a reason why a moving target appears smeared in a stationary SAR image. The Doppler parameters of a moving target can be represented as

$$f_{dc} = \frac{2V_r}{\lambda} \tag{2.3.6a}$$

$$f_{dr} = \frac{2[(V_a - V_y)^2 - R_0 a_r]}{\lambda R_0} \tag{2.3.6b}$$

$$H_{rwm} = \exp\left(j\frac{4\pi K_r}{c}V_r t_r t_a\right) \tag{2.3.6c}$$

$$H_{rcm} = \exp\left(-j\frac{4\pi K_r}{c}\frac{(V_a - V_y)^2 - R_0 a_r}{2R_0}t_r t_a^2\right) \tag{2.3.6d}$$

where f_{dc} represents the Doppler centroid, H_{rwm} and H_{rcm} represent RWM and RCM, respectively.

From Eqs. 2.3.6a, 2.3.6b, 2.3.6c, 2.3.6d it can be noted that Doppler centroid and RWM are impacted by the cross-track velocity of a moving target, whereas the Doppler modulation rate and RCM is affected by both the along-track velocity and the cross-track acceleration.

Therefore, the Doppler parameters of a moving target changes correspondingly with the motion parameters of the moving target. The relationship between the Doppler parameters and the motion parameters is the reason why a moving target appears smeared and dislocation in the image, and also the lead to indicate and focus a moving target. All moving target processing algorithms are based on one or several parameters in Eqs. 2.3.6a, 2.3.6b, 2.3.6c, 2.3.6d, thus the impact of each parameter is analyzed as follows.

2.3.2 Doppler Centroid

The cross-track velocity is the most interesting parameter of the moving target in GMTI applications. Therefore, in most existing algorithms, the moving target is classified according to the values of their cross-track velocities [1–3]. For example, if the moving target has a slow cross-track velocity that its spectrum is submerged by the clutter, it is defined as a slow moving target. However, when the Doppler centroid of the moving target exceeds the limit of PRF, the Doppler ambiguity is induced. The Doppler ambiguity exists when the cross-track velocity meets the condition

$$V_x \geq \frac{\lambda \cdot PRF}{2} \tag{2.3.7}$$

In that case, the "fast" or "slow" cross-track velocity [4] cannot accurately reflect the spectrum character of the moving target. In the existence of the Doppler ambiguity, the moving targets can be classified according to the locations of their spectra. We classify the moving targets into three types, as shown in Fig. 2.4.

The red and black triangles in Fig. 2.4 represent the Doppler spectra of the moving target and the clutter, respectively. The spectra of the moving targets of Type I are completely located out of the clutter. Hence, a high-pass filter is able to separate the moving target from the clutter. In Type II, partial spectra of the moving

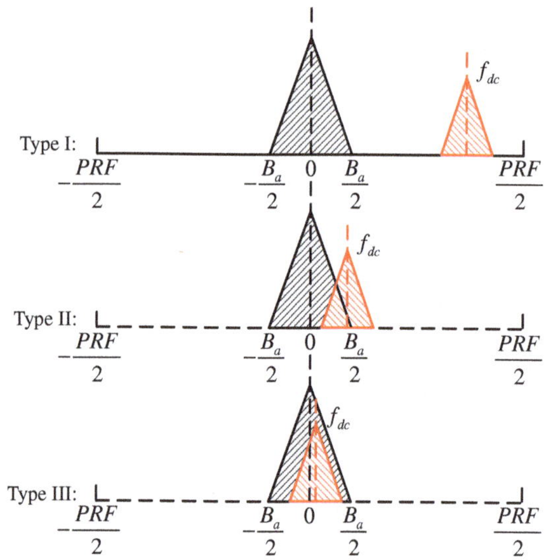

Fig. 2.4 Illustration of the three types of moving targets in the range-Doppler domain

target are submerged by the clutter. Type III represents the moving targets that are completely submerged by the clutter. The targets with slow cross-track velocities and the targets with fast cross-track velocities that aliased into the baseband, belong to Type III. In addition, given that the along-track velocity has no relationship with the Doppler centroid, the along-track moving targets also belong to Type III. The velocity information of the moving targets for each type is illustrated in Table 2.1, where B_a and B_m denote the Doppler bandwidths of the clutter and the moving target, respectively, and k is the number of the Doppler ambiguity.

In a whole synthetic aperture, the number of azimuth samples $N_a = PRF \cdot T_a$, where T_a represents the synthetic aperture time. Thus the number of azimuth dislocation samples can be calculated as

$$\Delta N = f_{dc}/PRF \cdot N_a \tag{2.3.8}$$

The synthetic aperture time $T_a = \frac{\lambda R_0}{L_a V_a}$, where L_a is the azimuth aperture size. Substitute it into Eq. 2.3.8, it yields that

$$\Delta N = \frac{2V_r R_0}{L_a V_a} \tag{2.3.9}$$

Table 2.1 Classification of moving targets

Type I	$\frac{B_a+B_m}{2} + k \cdot PRF < \lvert\frac{2V_r}{\lambda}\rvert < \frac{PRF}{2} + k \cdot PRF$
Type II	$\frac{B_a-B_m}{2} + k \cdot PRF < \lvert\frac{2V_r}{\lambda}\rvert < \frac{B_a+B_m}{2} + k \cdot PRF$
Type III	$k \cdot PRF \leq \lvert\frac{2V_r}{\lambda}\rvert < \frac{B_a-B_m}{2} + k \cdot PRF$

Substitute azimuth sample interval $d_a = \frac{L_a}{2}$ into Eq. 2.3.9, the azimuth dislocation is

$$\Delta R = \frac{V_r}{V_a} R_0 \tag{2.3.10}$$

According to Eq. 2.3.10, the azimuth dislocation of a moving target is determined by its cross-track velocity.

2.3.3 Doppler Modulation Rate

According to Eq. 2.3.6b, the Doppler modulation rate of a moving target is related to the along-track velocity and cross-track acceleration. The Doppler modulation rates of a moving target and a stationary target differ in the existence of these two motion parameters. The change of the Doppler modulation rate can be expressed as

$$\Delta f_{dr} = \frac{2V_a^2}{\lambda R_0} - \frac{2\left[(V_a - V_y)^2 - R_0 a_r\right]}{\lambda R_0} = \frac{2\left[V_y(2V_a - V_y) + R_0 a_r\right]}{\lambda R_0} \tag{2.3.11}$$

The Doppler modulation rate directly affects the azimuth focusing performance. In the existence of Δf_{dr}, a moving target is smeared since there is the second-order azimuth phase error. The second-order azimuth phase error will lead to the main-lobe broaden, the peak declination of the main-lobe and peak rise of the side-lobes. The second-order azimuth phase error of a moving target in a stationary SAR image can be expressed as

$$\Delta \phi_a = \exp\left\{ j \frac{2\pi}{\lambda R_0} [V_y(2V_a - V_y) + R_0 a_r] t_a^2 \right\} \tag{2.3.12}$$

Equation 2.3.12 reaches its largest value at the edge of the synthetic aperture, therefore the maximum value of the second-order azimuth phase error is

$$\Delta \varphi_a = \frac{\pi}{2\lambda R_0} \left[V_y(2V_a - V_y) + R_0 a_r\right] T_a^2 \tag{2.3.13}$$

By using the typical airborne and spaceborne SAR system parameters as shown in Table 2.2 [5], the numerical second-order phase error can be calculated. Suppose the along-track velocity of a moving target is 10 m/s and the cross-track acceleration is 0.1 m/s^2, the second-order azimuth phase error is 149.43 rad in the airborne SAR situation, while the value is 3.013 rad in the spaceborne situation. An airborne SAR has a smaller velocity and nearer range length, and has a higher resolution. Therefore, the defocus of a moving target is more obvious in an airborne SAR image than in a spaceborne SAR image.

Table 2.2 Typical system parameters in airborne and spaceborne SAR

Platform	System parameters					
	Wavelength (m)	Range (km)	Platform velocity (m/s)	Synthetic aperture time (S)	Azimuth resolution (m)	Range resolution (m)
Airborne	0.032	850	7100	3.4	0.5	1.25
Spaceborne	0.057	30	250	0.64	5	6.25

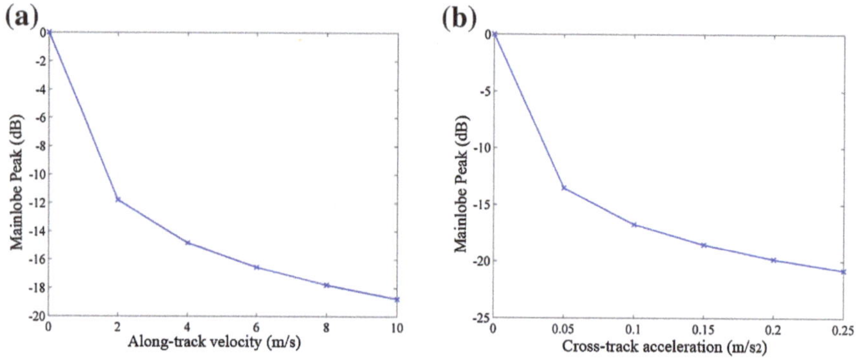

Fig. 2.5 Peak changes of the mainlobe with motions. **a** Changes with along-track velocity. **b** Changes with cross-track velocity

According to the analysis above, airborne SAR moving target imaging is a far more complicate issue than spaceborne SAR. Since the second-order azimuth error also causes the peak declination of the main-lobe, the moving target indication performance will be affected by using CFAR technique. By using the system parameters in Table 2.2, a set of simulations is operated, as shown in Fig. 2.5.

In Fig. 2.5a, the horizontal ordinate denotes the along-track velocity, which is 2, 4, 6, 8, 10 m/s; the vertical ordinate denotes the peak of the mainlobe, which is measured by dB. It can be noted that the peak of the mainlobe decreases with the increase of the along-track velocity. Figure 2.5b illustrates the relationship between the peak of the mainlobe and the cross-track acceleration. The cross-track accelerations are valued as 0.05, 0.1, 0.15, 0.2 and 0.25 m/s^2. It can be noted that the peak of the main-lobe also decreases with the increase of the cross-track acceleration. Therefore, it is proved that the change of Doppler modulation rate results in the smear of the image.

2.3.4 RCM

RCM is a phase error induced by the long synthetic aperture time. RCM reflects the changes of the instantaneous slant range, and exists in both the stationary and moving targets. The existence of RCM will lead to the energy of the target smear among multiple range gates, and will severely deteriorate the imaging quality.

According to the last exponential term of Eq. 2.2.5, in a broadside airborne SAR system, only the range curve migration exists. Thus, for the stationary target that locates at the center of the scene, its range curve migration reaches its largest value at the edge of the synthetic aperture. The largest range curvature can be expressed as

$$S_{rcm} = \frac{V_a^2}{2R_0}\left(\frac{T_a}{2}\right)^2 \tag{2.3.14}$$

Equations 2.3.6c and 2.3.6d are the RCM of a moving target. It can be noted that an additional RWM is induced by the cross-track velocity of the moving target, and the range curvature of a moving target is different from that of a stationary target. The range walk and curve migration of a moving target are expressed respectively as

$$S_{rwm} = V_r \frac{T_a}{2} \tag{2.3.15a}$$

$$S_{rcm} = \frac{(V_a - V_y)^2 - a_r R_0}{2R_0}\left(\frac{T_a}{2}\right)^2 \tag{2.3.15b}$$

The changes of RCM can be expressed as

$$\Delta S_{rwm} = V_r \frac{T_a}{2} \tag{2.3.16a}$$

$$\Delta S_{rcm} = \frac{2V_a V_y - V_y^2 + a_r R_0}{2R_0}\left(\frac{T_a}{2}\right)^2 \tag{2.3.16b}$$

According to the system parameters in Table 2.2, the values of Eq. 2.3.14 to Eqs. 2.3.15a, 2.3.15b can be calculated. Suppose the cross-track velocity of the moving target is 10 m/s, the cross-track acceleration is 0.1 m/s^2, and the along-track velocity is 10 m/s, in an airborne SAR situation, the range walk and curve migration of the moving target is 17 and 2.6299 m, and the range curve change is 0.3805 m. Compared with the range resolution 1.25 m, the RCM is far larger than the range resolution, which has to be corrected during imaging.

However, in the spaceborne SAR situation, the RCM of a stationary target is 3.0365 m, and the range walk and curve migration of the moving target is 3.2 and 3.0228 m. Compared with the range resolution 6.25 m, the RCM of the moving target is small, and the image resolution will not be significantly affected by the RCM of the moving target.

In conclusion, the RCM is much more severer in an airborne SAR system than in a spaceborne SAR system. Since the RCM is related to the motion parameters of a moving target, it is a very important parameter in GMTI and GMTIm algorithms.

2.4 Principle of Existing GMTI Algorithms

All existing GMTI algorithms use the Doppler characters of moving targets and their differences from those of the stationary targets to accomplish the target detection. There are mainly two parts in GMTI algorithms: clutter suppression and CFAR detection.

2.4.1 Principle of Single-Antenna GMTI Algorithms

Different GMTI algorithms use different methods to suppress the clutter. By suppressing the clutter, the spectra of moving targets can be separated from the clutter, and the SCR is increased.

(a) Spectrum filtering algorithm

According to the analysis in Sect. 2.3.2, a Doppler centroid shift is induced by the cross-track velocity of a moving target. The spectrum of the moving target will move towards the high-band of the PRF with the increase of the cross-track velocity, and the spectrum of the target will be eventually located outside the clutter. Based on the Doppler centroid shift, the moving target can be detected by setting up a series of band-pass filter in the high-band of PRF. Spectrum within the band-pass filter will be restrained, and the clutter will be suppressed. After the clutter suppression, the spectrum of the moving target can be extracted, and the target can be re-focused, as shown in Fig. 2.6.

In Fig. 2.6, n band-pass filters with different frequency center are set in the range-Doppler domain, and the clutter is not located within the band-pass filters. If the target has only the cross-track velocity, the target can be focused after shifting its spectrum into baseband. The flowchart of spectrum filtering algorithm is illustrated in Fig. 2.7.

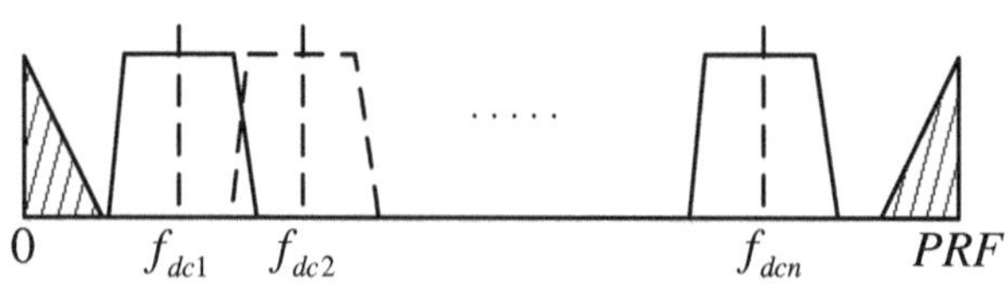

Fig. 2.6 Principle of the spectrum filtering algorithm

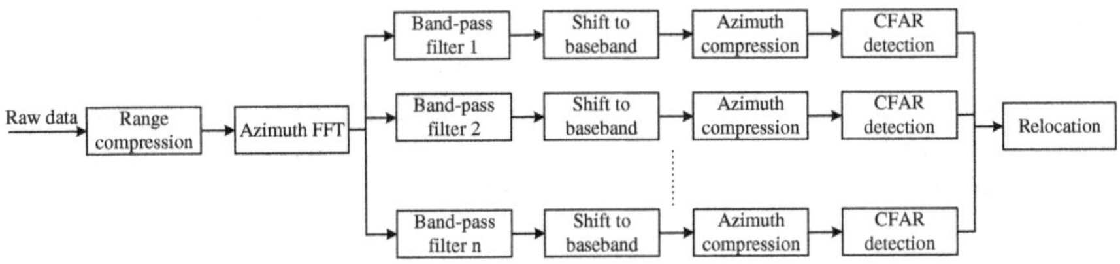

Fig. 2.7 Flowchart of the spectrum filtering algorithm

However, the spectrum filtering algorithm requires that the cross-track velocity of the moving target will not induce significant RCM, i.e., the RCM of the target must be less than half of the range resolution.

$$\left|\frac{V_r}{V_a}\right| < \frac{1}{\beta}\frac{\rho_r}{2R_0} \tag{2.4.1}$$

where β is the beam angle, ρ_r denotes the range resolution.

Moreover, the spectrum filtering algorithm requires that there is no Doppler ambiguity. The cross-track velocity has to satisfy the condition that

$$\frac{1}{2}\mu\beta < \left|\frac{V_r}{V_a}\right| < \frac{\lambda \cdot PRF}{2V_a} - \frac{\beta}{2} \tag{2.4.2}$$

where μ is the coefficient of antenna and signal processing.

The Doppler ambiguity will be induced in most practical applications. However, although the spectrum filtering algorithm requires that no Doppler ambiguity exists, it can still suppress the clutter only if the spectrum of the moving target locates outside the clutter.

In conclusion, there are mainly three limitations of spectrum filtering algorithm: first, the velocity detection range has a inconsistent requirement with the observation swath; second, the spectrum filtering algorithm cannot detect targets with slow cross-track velocity and blind velocity; third, targets with only along-track velocities cannot be detected.

(b) RDM algorithm

RDM algorithm is a representative algorithm that detects moving targets with the Doppler modulation rate change. It is firstly invented as a motion compensation algorithm that corrects the motion error of the platform during the flight. J.R. Moreira first applied this algorithm in GMTI, the principle of RDM algorithm is: The change of Doppler modulation rate can be detected by the relationship between neighboring Doppler power spectra. The moving target can be indicated by its Doppler centroid, and the along-track velocity and cross-track acceleration can be estimated.

There are mainly three limitations of RDM algorithm: first, only targets with the along-track velocity and cross-track acceleration can be detected, but the target with only the cross-track velocity cannot be detected; second, RDM algorithm is not effective for the detection of multiple targets which are located nearby; third, the Doppler parameter estimation precision is not high, and two motion parameters cannot be derived from one Doppler parameter.

Shear average algorithm also uses the Doppler modulation rate to detect moving targets. This algorithm uses the contrast ratio of sub-images focused with different Doppler modulated rates to judge whether there is a moving target in the sub-image. Moreover, shear average algorithm has also the disadvantages same as RDM algorithm.

(c) Time-frequency analysis algorithm

The signal character differences between a moving target and a stationary target can be revealed in the time-frequency domain. Motions of a moving target changes the Doppler centroid and Doppler modulation rate of the moving target, and time-frequency analysis algorithms are highly efficient in extracting the motion parameters. WVD transform is proposed by S. Barbarossa to extract and estimate the motion parameters of moving targets. After WVD transform, the azimuth spectrum is transformed into a straight line, the slope of the line is the Doppler modulation rate, and the frequency initial is the Doppler centroid. Suppose the signal is $s(t)$, the definition of WVD transform is

$$W(t,f) = \int s\left(t+\frac{\tau}{2}\right)s^*\left(t-\frac{\tau}{2}\right)\exp(-j2\pi\tau f)d\tau \tag{2.4.3}$$

The WVD transform has a poor performance in the low SNR condition, so the clutter suppression and spectrum extraction must be performed before WVD. Thus, WVD transform is more suited as a GMTIm algorithm but not a GMTI algorithm.

Furthermore, WVD transform is a bilinear transform, and is not suitable for multi-target detection and parameter estimation. Improved WVD transforms have been studied, while the performances are not promising as well. Wavelet transform is also a sort of new time-frequency analysis algorithms, but it has a heavy calculation burden and not suited for practical applications.

2.4.2 CFAR Detection

CFAR detection is a classical technique of radar detection. In a radar system, if the detection threshold V_T is set as a constant value, the false-alarm rate P_{fa} will rise with the increase of interference σ:

$$P_{fa} = \exp\left(-\frac{V_T^2}{2\sigma^2}\right) \tag{2.4.4}$$

In a radar system, there are not only the thermal noise within the receiver, but also the interferences from the clutter and the enemy. These interferences are strong in most situations, and their amplitudes vary in different scenes. In that case, the false-alarm rate will not be constant any more. The method to keep a constant false-alarm rate is called CFAR detection [6].

With the development of radar detection, many CFAR detection methods are invented [7–13]. In practical applications, the specific method must be chosen according to the echo distribution. In a low-resolution SAR system, the amplitude of the echo obeys the Rayleigh distribution, while the amplitude of the echo obeys other distributions such as Swerling model [14], lognormal model [15] and χ^2 distribution. In most SAR data, the Rayleigh distribution assumption is convincing, so the classical CA-CFAR algorithm in Rayleigh distribution is introduced in this section.

Suppose the probability distribution function of clutter y is expressed as

$$p(y) = \frac{y}{\sigma^2}\exp\left(-\frac{y^2}{2\sigma^2}\right) \tag{2.4.5}$$

Let $x = \frac{y}{\sigma}$, $V_0 = \frac{V_T}{\sigma}$, and substitute them into Eq. 2.4.4, it yields

$$P_{fa} = \exp\left(-\frac{V_0^2}{2}\right) \tag{2.4.6}$$

For the new clutter, the false-alarm rate follows Eq. 2.4.6, which is not related to the clutter. The relationship between the derivation σ and the mean μ is

$$\sigma = \sqrt{\frac{2}{\pi}}\mu \tag{2.4.7}$$

The principle of CA-CFAR is illustrated in Fig. 2.8.

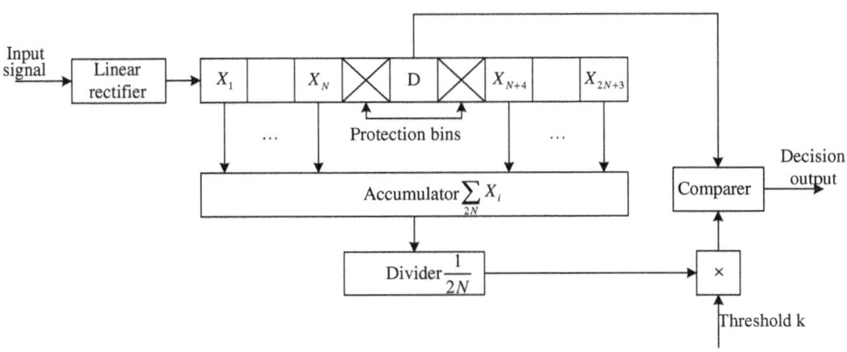

Fig. 2.8 Principle of CA-CFAR detection

2.5 Principle of GMTIm Algorithms

After GMTI, the echoes of moving targets are separated from that of the clutter. GMTIm algorithms are used to focus and relocated the moving target by estimating its Doppler parameters. The main operations of GMTIm algorithms contain RCMC and Doppler parameter estimation.

2.5.1 *RCMC of Moving Targets*

The RCM of the moving target must be accurately corrected so as to get a focused image of the target. However, the RCM of the moving target is different from that of the clutter, and is unknown without the information of motion parameters. In most articles, the RCMC of moving targets is performed by using the Keystone transform [16].

Before the range compression, the echo of a point target in a broadside airborne SAR can be expressed as

$$s_r(t_a, t_r) = A_0\omega_a(t_a)rect\left[\frac{t_r - \frac{2R(t_a)}{c}}{T_r}\right]\exp\left(j\pi K_r\left(t_r - \frac{2R(t_a)}{c}\right)^2\right)\exp\left(-j\frac{4\pi f_c}{c}R(t_a)\right) \tag{2.5.1}$$

where T_r denotes the pulse duration, f_c denotes the carrier frequency. Transform Eq. 2.5.1 into range-frequency domain, it yields

$$s_r(t_a, f_r) = A_0\omega_a(t_a)rect\left[\frac{f_r}{K_r T_r}\right]\exp\left(-j\pi\frac{f_r^2}{K_r}\right)\exp\left(-j\frac{4\pi(f_c+f_r)}{c}R(t_a)\right) \tag{2.5.2}$$

where f_r is the range frequency. In Eq. 2.5.2, the first exponential term can be compensated by a range matched filter or the deramping operation so as to accomplish the range compression. The RCM and azimuth phase information are hidden in the second exponential term. Suppose it is represented by $\Phi(t_a, f_r)$, substitute Eq. 2.3.4 into Eq. 2.5.2, it yields

$$\begin{aligned}\Phi(t_a, f_r) = {} & \exp\left(-j\frac{4\pi(f_c+f_r)}{c}R_0\right)\exp\left(j\frac{4\pi V_r}{c}(f_c+f_r)t_a\right) \\ & \cdot\exp\left(-j\frac{2\pi((V_a - V_y)^2 - a_r R_0)}{cR_0}(f_c+f_r)t_a^2\right)\end{aligned} \tag{2.5.3}$$

In Eq. 2.5.3, the second exponential term contains the Doppler centroid shift and the RWM; the third exponential term contains the second-order azimuth phase error and the range curve migration. It is noted that the coupling of the range-frequency and azimuth-time denotes the RWM, and the coupling of the range-frequency and the second-order azimuth-time denotes the range curve migration. Transform the azimuth time t_a by

$$t_a = \frac{f_c}{f_c + f_r}\tau_a \tag{2.5.4}$$

Substitute it into Eq. 2.5.3, it yields

$$\begin{aligned}\Phi(t_a, f_r) &= \exp\left(-j\frac{4\pi(f_c + f_r)}{c}R_0\right)\exp\left(j\frac{4\pi V_r}{c}f_c\tau_a\right)\\ &\cdot \exp\left(-j\frac{2\pi((V_a - V_y)^2 - a_r R_0)}{cR_0}\frac{f_c^2}{f_c + f_r}\tau_a^2\right)\end{aligned} \tag{2.5.5}$$

It is noted that the range-frequency and azimuth-time is decoupled after the transform, and in turn the RWM is corrected. Equation 2.5.4 is the definition of Keystone transform [17–19]. By using the Keystone transform, the RWM can be corrected without the estimation of the cross-track velocity V_r.

It is also noted from Eq. 2.5.5 that the range curve migration is not corrected by using the Keystone transform. Thus, researchers proposed the second-order Keystone transform [20], which is defined as

$$t_a = \left(\frac{f_c}{f_c + f_r}\right)^{\frac{1}{2}}\tau_a \tag{2.5.6}$$

Substitute Eq. 2.5.6 into Eq. 2.5.3, it yields

$$\begin{aligned}\Phi(t_a, f_r) &= \exp\left(-j\frac{4\pi(f_c + f_r)}{c}R_0\right)\exp\left(j\frac{4\pi V_r}{c}f_c^{\frac{1}{2}}(f_c + f_r)^{\frac{1}{2}}\tau_a\right)\\ &\cdot \exp\left(-j\frac{2\pi((V_a - V_y)^2 - a_r R_0)}{cR_0}f_c\tau_a^2\right)\end{aligned} \tag{2.5.7}$$

It can be noted that the range curve migration is corrected after the second-order Keystone transform, and half of the RWM is removed. In the moving target imaging of an airborne SAR system, the impact of the RWM on the imaging quality of a moving target is far severer than the range curve migration [21], so the first-order Keystone transform is the most widely-used algorithm in RCMC of moving targets, and the second-order Keystone transform is only suited for the imaging of along-track moving targets.

Since the azimuth time is a discrete value, the Keystone transform is accomplished by sinc interpolation. The precision of sinc interpolation is highly related to the length of the interpolation core, and long interpolation core requires heavy calculation burden. Therefore, Keystone transform is not suitable for real-time processing.

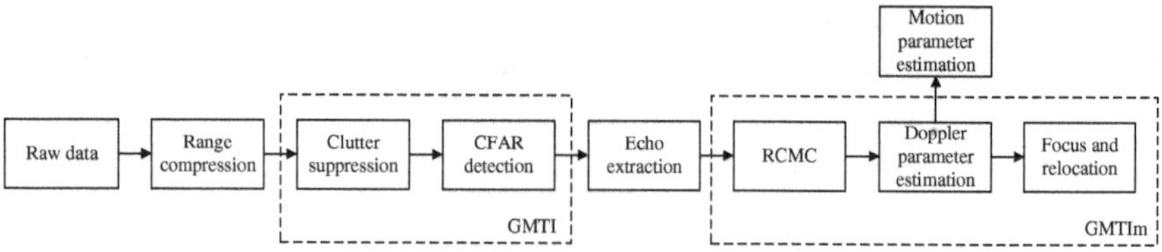

Fig. 2.9 Flowchart of GMTI and GMTIm in airborne SAR

Another disadvantage of Keystone transform is that it is only effective without the existence of Doppler centroid ambiguity. If there is Doppler centroid ambiguity in the echo, the Doppler centroid ambiguity number must be estimated before Keystone transform.

2.5.2 *Motion Parameter Estimation*

After RCMC, the energy of the moving target is concentrated into a range gate, and the echo of the moving target can be viewed as a 1-D signal. The Doppler parameters are estimated from the azimuth signal, and the target can be focused and relocated using the estimations.

Since the azimuth signal of the target in SAR can be viewed as a chirp signal, the Doppler parameters can be accurately estimated using the time-frequency analysis algorithms, such as WVD transform, FrFT [22], wavelet transform and so forth.

After the Doppler parameter estimation, motion parameters can be derived. The cross-track velocity can be calculated as

$$\hat{V}_r = \frac{\lambda \hat{f}_{dc}}{2} \tag{2.5.8}$$

Since the two motion parameters cannot be derived from one Doppler parameter, the cross-track acceleration and the along-track velocity cannot be derived from the Doppler modulation rate at the same time. Thus, the acceleration is neglected in practical applications. The along-track velocity is calculated as

$$\widehat{V}_y = V_a - \sqrt{\frac{\lambda R_0 \hat{f}_{dr}}{2}} \tag{2.5.9}$$

If there are multiple moving targets in the scene, each target must be estimated and focused independently. The flowchart of the whole moving target processing is illustrated in Fig. 2.9.

2.6 Conclusion

The principle of SAR imaging has been briefly introduced in this chapter, and the echo signals of stationary and moving targets in a broadside airborne SAR system have been established. By comparing the Doppler parameters between the stationary and moving targets, the relationship between motion parameters of moving targets and the Doppler centroid, Doppler modulation rate and RCM has been analyzed.

Based on the analysis, three classical single-channel GMTI algorithms have been introduced in this chapter, and their limitations have been discussed. Moreover, the principle of GMTIm algorithms with Keystone transform and time-frequency analysis algorithm has been introduced, and the limitations in Doppler centroid ambiguity and motion parameter derivation have been studied.

The content in this chapter is the foundation of the following chapters. The limitations of existing algorithms mentioned in this chapter are further studied and solved in the proposed algorithms in this book.

References

1. Qian J, Lv X, Xing M et al (2010) Parameter estimation of the fast moving target based on tri-channel airborne SAR-GMTI. J Xidian Univ 37(2):235–241
2. Dai D, Liu J, Tang H et al (2005) Slowly moving target detection and imaging using single-antenna SAR. Sig Process 21(4A):550–554
3. Zhu S, Liao G, Zhou Z et al (2009) Approach to ground slowly moving target parameter-estimation for airborne dual-channel SAR systems. Syst Eng Electron 31 (12):2848–2852
4. Zhao Y (2007) Ground moving target indication of synthetic aperture radar. Institute of Electronics, Chinese Academy of Sciences
5. Cumming IG, Wong FH (2005) Digital processing of synthetic aperture radar data: algorithms and implementation. Artech House, Norwood
6. Nitzberg R (1972) Constant-False-Alarm-Rate signal processors for several types of interference. IEEE Trans Aerosp Electron Syst 8(1):27–34
7. He Y, Guan J, Peng Y et al (1999) Radar target detection and CFAR technique. Tsinghua University Press, Beijing
8. Rohling H (1983) Radar CFAR thresholding in clutter and multiple target situations. IEEE Trans Aerosp Electron Syst 19(3):608–621
9. Ji S, Huangpu K, Zhou L et al (1990) A new constant-false-alarm-rate (CFAR) processor. J Nat Univ Defense Technol 12(4):116–121
10. Thomas JB (1970) Nonparametric detection. Proc IEEE 58(5):623–631
11. Guan J, He Y, Peng Y (2000) Study of distributed CFAR detection based on local test statistics and its fusion strategy. ACTA Electronica Sin 28(3):125–127
12. Cai LJ, Wang H (1994) Further results on adaptive filtering with embedded CFAR. IEEE Trans Aerosp Electron Syst 30(4):1009–1020
13. Wanielik G, Stock DJR (1990) Measured scattering-matrix-data and a polarimetric CFAR-detector which works on this data. In: IEEE international radar conference, Piscataway, NJ, USA, 1990, pp 514–519

14. Swerling P (1997) Radar probability of detection for some additional fluctuating target cases. IEEE Trans Aerosp Electron Syst 33(2):698–709
15. Trunk GV, George SF (1970) Detection of targets in non-Gaussian sea clutter. IEEE Trans Aerosp Electron Syst 6(5):620–628
16. Perry RP, Dipietro RC, Fante RL (1999) SAR imaging of moving targets. IEEE Trans Aerosp Electron Syst 35(1):188–200
17. Tang L, Li D, Wu Y et al (2005) Imaging of ground moving targets based on airborne SAR. Syst Eng Electron 27(10):1681–1684
18. Qian J, Su J, Li L et al (2010) Tri-channel SAR-GMTI high-speed target imaging and motion parameter estimation using KWT. J Electron Inf Technol 32(7):1660–1667
19. Yin J, Li D, Tang L et al (2007) Spaceborne SAR detecting and imaging of airborne moving targets. ACTA Electronica Sin 35(6):1032–1036
20. Zhou F, Li Y, Xing M et al (2007) An effective approach to ground moving target imaging and motion parameter estimation for single channel SAR system. ACTA Electronica Sin 35 (3):543–548
21. Yang J, Huang X, Jin T et al (2011) New approach for SAR imaging of ground moving targets based on a Keystone transform. IEEE Geosci Remote Sens Lett 8(4):829–833
22. Sun H, Liu G, Gu H et al (2002) Application of the fractional Fourier transform of moving target detection in airborne SAR. IEEE Trans Aerosp Electron Syst 38(4):1416–1424

Chapter 3
Fast Moving Target Indication and Imaging in Stripmap SAR

Abstract This chapter mainly discusses the GMTI and GMTIm algorithms in stripmap SAR/GMTI mode. The signal model of fast moving targets in stripmap SAR/GMTI mode is established, and the impacts of fast cross-track velocities on the Doppler ambiguity and higher order azimuth phase are analyzed. Furthermore, a classification of targets by the locations of their spectra is presented. Based on the classification, a Doppler centroid estimation algorithm based on curve fitting, a multiple target indication and echo extraction method, and a fast moving target imaging algorithm based on Hough transform and third-order PFT are presented. Finally, simulations and real data are utilized to prove the effectiveness of these algorithms.

3.1 Introduction

The stripmap mode is a fundamental and most widely-used working mode of the SAR system. High-resolution images of the observation area can be stably obtained in the stripmap mode since the antenna is fixed in this mode. Compared with the spotlight mode, the stripmap mode has a broader observation area, while compared with the scan-SAR mode, it has a higher resolution. Along with the development of the moving target processing technique, both the ability of moving target processing and the high resolution imaging are required in the stripmap SAR, therefore this new mode is called SAR/GMTI mode.

Chapter 2 of this book induced the broadside geometry of a moving target in the stripmap mode, and also some classical GMTI and GMTIm algorithms are introduced in Sects. 2.4 and 2.5. These algorithms are effective for the targets with certain scopes of motion parameters. As analyzed in Chapter Two, in the existence of the target with complex motions, such as Doppler ambiguity, these algorithms cannot accurately indicate, estimate and focus the moving target. In the real SAR signal processing, the motions of the moving target is unknown, the possibility of the existence of Doppler ambiguity cannot be ignored, therefore it is necessary to develop the fast moving target indication and imaging algorithms.

J. Yang, *Study on Ground Moving Target Indication and Imaging Technique of Airborne SAR*, Springer Theses, DOI 10.1007/978-981-10-3075-8_3

In order to solve the fast moving target indication and imaging in stripmap SAR, an adaptive Doppler centroid estimation algorithm is introduced in this chapter. By using the curve fitting of the azimuth spectrum, the Doppler centroid can be accurately estimated in the non-homogeneous scene, and the impact of the prominent targets is eliminated. With the accurate estimation of the Doppler centroid, the Doppler centroid of the stationary scene can be adjusted into zero, which is the foundation of the following moving target processing algorithms.

After adjusting the Doppler centroid error of the clutter, a multiple moving target indication and echo extraction method is proposed. This proposed two-step algorithm indicates the moving targets with different motion parameters based on the classification in Sect. 2.3.2, which can indicate the fast moving target submerged by the clutter. After GMTI, the impacts of the integrity and SNR of the echo extraction on the following parameter estimation and imaging algorithms are discussed.

Then, a fast moving target imaging algorithm based on Hough transform and third-order PFT is presented. This algorithm can accurately estimate the real Doppler centroid in the existence of the Doppler ambiguity. And the third-order azimuth phase error is compensated by using the third-order PFT. The effectiveness of the algorithm is sufficiently proved by simulations and real data processing.

This chapter is organized as follows. Based on Sect. 2.3, the echo model of moving targets with fast cross-track velocity is established in Sect. 3.2. An adaptive Doppler centroid estimation algorithm is introduced in Sect. 3.3, and a two-step multiple moving target indication and echo extraction method is proposed in Sect. 3.4. Then, a fast moving target imaging algorithm based on Hough transform and third-order PFT is introduced in Sect. 3.5. Finally, conclusive remarks are provided in Sect. 3.6.

3.2 Echo Model of Fast Moving Target

The echo model of fast moving target is established in this section. No matter what the motion parameters of the target are, the geometry of the moving target in broadside SAR is unchanged. Therefore, the echo model is established based on Figs. 2.2 and 2.3.

According to Eq. 2.3.4, cross-track velocity, cross-track acceleration and along-track velocity are the main parameters that affect the moving target imaging. In fact, the impact of the along-track acceleration is existed. However, its impact is neglected for the following reasons: first of all, the Doppler centroid and Doppler modulation rate are not related to the along-track acceleration; secondly, in the real flight of airborne SAR, the platform velocity is not stable, and the along-track acceleration of the moving target is negligible compared with the changes of the platform velocity, and can be compensated during the autofocus process. Thus, the impact of the along-track acceleration is neglected, and the instantaneous slant range of the moving target can be expressed as

$$R(t_a) = \sqrt{H^2 + \left(X - V_x t_a - \frac{1}{2} a_x t_a^2\right)^2 + (V_a t_a - V_y t_a)^2} \tag{3.2.1}$$

Expand Eq. 3.2.1 into a Taylor series and keep it to the third-order term of t_a, then Eq. 3.2.1 can be approximated as

$$R(t_a) \approx R_0 - \frac{X}{R_0} V_x t_a + \frac{(V_a - V_y)^2 - a_x X}{2R_0} t_a^2 + \frac{X V_x (V_a - V_y)^2}{2R_0^2} t_a^3 \tag{3.2.2}$$

Substitute Eq. 2.3.3 into Eq. 3.2.2, it yields

$$R(t_a) \approx R_0 - V_r t_a + \frac{(V_a - V_y)^2 - a_r R_0}{2R_0} t_a^2 + \frac{V_r (V_a - V_y)^2}{2R_0^2} t_a^3 \tag{3.2.3}$$

According to Eq. 3.2.3, the third-order Taylor expansion term of a moving target is related to the cross-track velocity and the along-track velocity, which increases accordingly with the cross-track velocity, and he third-order term is no longer negligible if the cross-track V_r is fast. Substitute Eq. 3.2.3 into Eq. 2.2.4, it yields

$$\begin{aligned} s_r(t_a, t_r) = {} & A_0 \omega_a(t_a) \operatorname{sin c}\left[\pi B_r \left(t_r - \frac{2R_0}{c}\right)\right] \exp\left(-j\frac{4\pi}{\lambda} R_0\right) \exp\left(j\frac{4\pi}{\lambda} V_r t_a\right) \\ & \cdot \exp\left(-j\frac{4\pi}{\lambda} \frac{(V_a - V_y)^2 - R_0 a_r}{2R_0} t_a^2\right) \exp\left(j\frac{4\pi K_r}{c} V_r t_r t_a\right) \\ & \cdot \exp\left(-j\frac{4\pi K_r}{c} \frac{(V_a - V_y)^2 - R_0 a_r}{2R_0} t_r t_a^2\right) \exp\left(-j\frac{4\pi}{\lambda} \frac{V_r (V_a - V_y)^2}{2R_0^2} t_a^3\right) \end{aligned} \tag{3.2.4}$$

The second exponential term of Eq. 3.2.4 is the Doppler centroid term of a moving target. If the cross-track velocity satisfies Eq. 2.3.7, the Doppler ambiguity will be induced. Assuming a Ku-band airborne SAR system with PRF of 1500 Hz, carrier frequency of 0.0194 m, then the Doppler ambiguity will be induced if the cross-track velocity of a moving target is faster than 14.55 m/s. In real applications, both civil and military vehicles can travel above this speed, which indicates that Doppler ambiguity cannot be neglected in the real data moving target processing.

The fourth exponential term is the RWM term of the moving target. It can be noted that the RWM significantly increases in the case of fast moving targets. The RWM and Doppler centroid are both induced by the cross-track velocity. The Doppler centroid is sampled by PRF, which is aliased with a period of PRF. However, the RWM is not affected by the Doppler ambiguity. Therefore, the fast cross-track velocity can be extracted from the RWM.

Compare Eq. 3.2.4 with Eq. 2.3.5, the only difference is that there is a third-order phase error in Eq. 3.2.4, i.e., the last exponential term in Eq. 3.2.4. It can be represented as

$$H_3 = \exp\left(-\mathrm{j}\frac{4\pi}{\lambda}\frac{V_r(V_a - V_y)^2}{2R_0^2}t_a^3\right) \tag{3.2.5}$$

It can be noted that the third-order phase error increases with the cross-track velocity. The third-order phase error will lead to the asymmetry of the side-lobes, and is not negligible for the fast moving target. Therefore, the third-order phase error must be compensated during the fast moving target imaging.

We simulated a single-antenna SAR system (see Table 3.1) to validate the conclusions above. In the observation area, four targets are set as shown in Table 3.2. It can be noted that T0 is a stationary target, T1 and T2 are slow moving target with a small cross-track velocity, and T3 is a fast moving target.

The processing results of the four targets are shown in Fig. 3.1. The images after the range compression are shown in Fig. 3.1a. It can be noted that the RWM curve of T0 are paralleled with the azimuth axis, while the RWM curves of T1, T2 and T3 are slant in different scale. The focusing results of these targets are shown in Fig. 3.1b by using the stationary target parameters. It can be noted that all targets are smeared excepted T0. Also, T1 and T2 are dis-located from their real azimuth location in the image with the existence of the cross-track velocity. The dis-location degree of T2 is larger than that of T1 since T2 has a larger cross-track velocity. However, T3 appears in the upper half of the image, which is induced by the Doppler ambiguity. This result proves that the Doppler centroid cannot be correctly estimated by using the azimuth location of the fast moving target.

Figure 3.2a, b compare the azimuth compression results of T1 and T3 with and without third-order phase error compensation. In the simulation, assuming that the RCM and azimuth second-order phase error are accurately corrected, it is noted that the impact of third-order phase error is negligible in the case of T1. On the contrary, the third-order phase error induces side-lobe symmetry of fast moving target T3.

According to the echo model and analysis in this section, three conclusions can be drawn: first of all, the large cross-track velocity of a fast moving target leads to the Doppler ambiguity; secondly, the Doppler ambiguity leads to the azimuth location aliasing, while the RWM is not affected; thirdly, the third-order azimuth phase error must be compensated during the imaging process.

Table 3.1 System parameters of the simulation

System parameter	Value	System parameter	Value
Central slant range	1000 m	Range sampling rate	60 MHz
Carrier frequency	2 GHz	PRF	400 Hz
Pulse time width	5 us	Platform velocity	100 m/s
Pulse bandwidth	30 MHz	Antenna size	1 m

Table 3.2 Target parameters of the simulation

Target	Motion parameter			
	Nearest slant range (m)	Cross-track velocity (m/s)	Along-track velocity (m/s)	Cross-track acceleration (m/s^2)
T0	800	0	0	0
T1	900	1	3	0.1
T2	1000	3	3	0.1
T3	1200	18	3	0.1

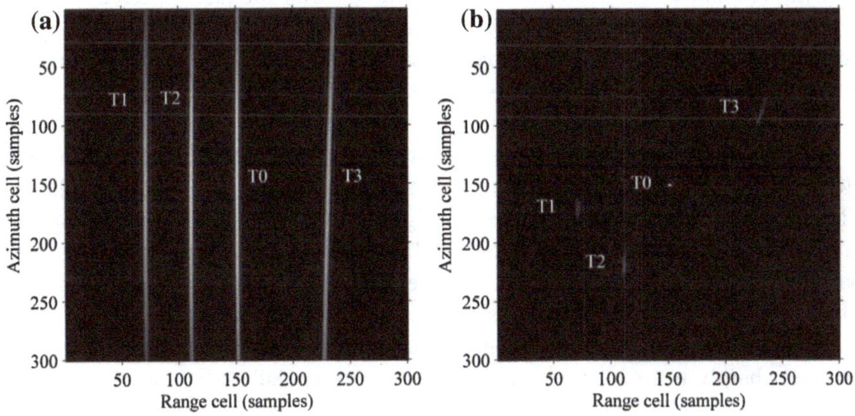

Fig. 3.1 Simulation results of the targets. **a** Range compression result. **b** Azimuth compression result

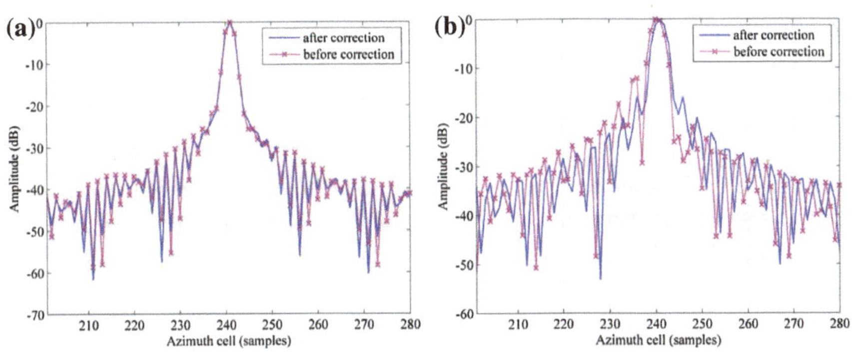

Fig. 3.2 Azimuth compression results of T1 and T3. **a** Azimuth compression result of T1. **b** Azimuth compression result of T3

3.3 Doppler Centroid Estimation of the Clutter

In our signal model establishment of stripmap SAR, the system is assumed to be with a squint angle of zero. However, in the real flight of an airborne SAR, because of the instability of the platform and the installation error of the antenna, there will be an antenna direction error. The antenna direction error induces a spectrum shift of the clutter, which will affect the azimuth window function and the RWM correction. Therefore, before the SAR imaging algorithm is preceded, the actual Doppler centroid of the clutter must be estimated.

Meanwhile, estimation and correction of the Doppler centroid of the clutter are also the foundation of GMTI in the azimuth spectrum domain in Sect. 3.4. The spectrum shift of the clutter will deteriorate the performance of the clutter suppression. Also, estimation and correction of the Doppler centroid of the clutter is important for the cross-track velocity estimation, which will be introduced in Sect. 3.5.

In a radar system, the Doppler centroid data is recorded by the INS system. However, the precision of the INS system cannot satisfy the imaging resolution. Therefore, Doppler centroid estimation algorithms based on the echo signal are produced [1]. Classical Doppler centroid estimation algorithms include energy balancing method [2], correlation Doppler estimator, SDE [3], and MLE [4]. Energy balancing method, which is widely used in airborne SAR imaging, has a promising performance in homogeneous scenes, whereas its estimation precision declines sharply if the homogeneity is destroyed, especially when prominent point targets exist in the scene.

3.3.1 *Principle of Energy Balancing Method*

The basis of energy balancing method is regarding that the amplitude of azimuth spectrum is generally the same as the pattern of antenna [5]. Thus, the energy center of azimuth spectrum can represent the estimation of Doppler centroid.

The energy balancing method is proceeded by cyclic iteration. First of all, the initial value of Doppler centroid is set, and the Doppler centroid is calculated by iterative solving the energy of the both sides of the Doppler centroid. The principle of the energy balancing method is shown in Fig. 3.3. Specific processing steps are as follows.

(a) After the range compression of the raw data, transform the data into azimuth frequency domain;
(b) Set the initial Doppler centroid f_{dc}^{0} by using the system parameters and the INS data;

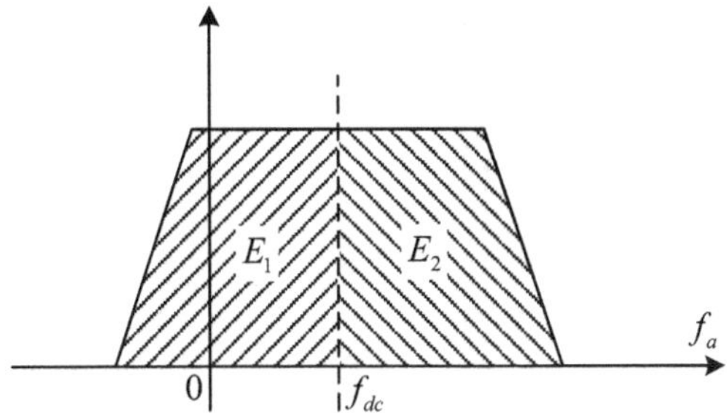

Fig. 3.3 Principle of the energy balancing method

(c) Calculate the energy of both sides E_1 and E_2, and calculate the normalized energy difference $\Delta E = \frac{E_1 - E_2}{E_1 + E_2}$;

(d) Solve the relation coefficient $\varepsilon = \left|\frac{\partial \Delta E}{\partial f_{dc}}\right|_{\Delta E = 0}$;

(e) Set $f_{dc}^{l+1} = f_{dc}^{l} + \Delta f_{dc}$, where l is an integer. Repeat the cycle, until $\Delta E = 0$. Then, the Doppler centroid estimation is $\hat{f}_{dc}$ solved.

Energy balancing method can provide an accurate Doppler centroid estimation in the homogeneous scene. However, if the scene is non-homogeneous, especially if prominent point targets exist in the scene, energy center cannot correctly replace Doppler centroid, and thus the energy balancing method is ineffective. One way to suppress the impact of the prominent point targets by using the average of multiple range gates. However, this method cannot eliminate the prominent point target, and the Doppler centroid relationship among different range gates is neglected. In [6], an improved energy balancing method is proposed by setting a threshold. This method can only eliminate targets which have a high energy peak than the threshold, and the continuity of the azimuth spectrum is destroyed.

3.3.2 *Adaptive Doppler Centroid Estimation Algorithm Based on Curve Fitting*

In order to improve the performance of the energy balancing method in the non-homogeneous scene, the prominent point targets must be eliminated. Figure 3.4 shows the azimuth power spectrum of real airborne SAR system in non-homogeneous scene. According to Fig. 3.4, there are mainly two kinds of prominent point targets that deteriorate the homogeneity of the scene. Target A represents targets with extremely high power that severely affects energy distribution of the spectrum, and Target B represents targets with considerable power, but relatively lower peak.

In [6], an algorithm is proposed to remove the impact of Target A by setting up rigid threshold. However, the value of threshold is given manually, and the impact of Target B is neglected.

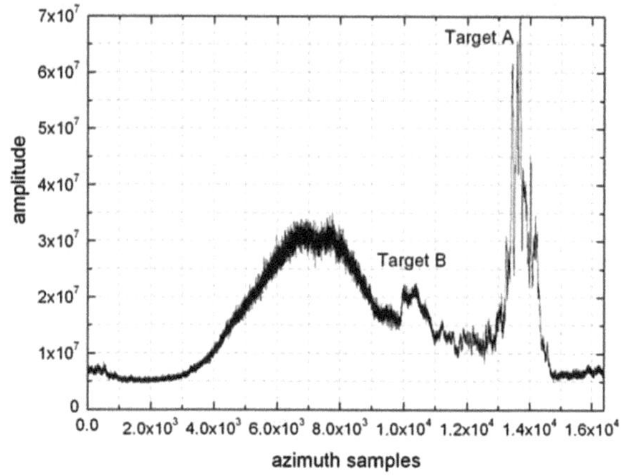

Fig. 3.4 Azimuth power spectrum of airborne SAR

According to the shapes of power spectra of prominent point targets, an improved energy balancing algorithm based on Gaussian curve fitting of down-sampled power spectrum is proposed. In this section, the principle and specific processing steps of the proposed algorithm are introduced.

Prominent point targets, such as Target A and B can be considered as disturbances of the Gaussian-shape azimuth spectrum. Moreover, their spectra are keen-edged and narrow, whereas the azimuth spectrum is relatively flat. After operate down-sample to the azimuth spectrum, only a few samples are located in Target A and B. On the contrary, most of the samples obey Gaussian distribution. Therefore, after Gaussian curve fitting, samples of ideal azimuth spectrum are preserved, and samples of prominent point targets will be abandoned adaptively.

The proposed algorithm consists of six steps as follows:

Step 1: Operating down-sample of the azimuth spectrum to get the discrete spectrum data;

Step 2: Operating Gaussian curve fitting of the discrete azimuth spectrum, thus samples of prominent point targets will be far from the fitting curve, which is defined as exceptional samples;

Step 3: Calculating the fitting errors of each sample, and calculate the mean and standard deviation of the fitting errors. Compare the fitting error of each sample with the sum of mean and standard deviation, the samples with fitting errors larger than the sum are regarded as exceptional samples and should be abandoned;

Step 4: Operating Gaussian curve fitting of the new spectrum without exceptive samples, the new fitting curve is considered to be the ideal azimuth power spectrum;

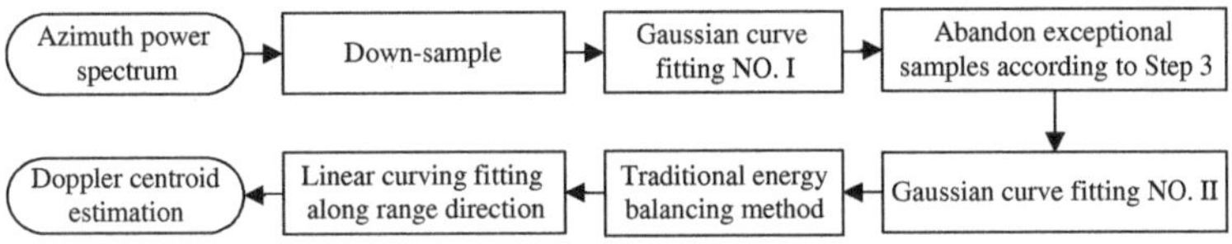

Fig. 3.5 Flowchart of the proposed algorithm

Step 5: Processing traditional energy balancing method to estimate Doppler centroid $\hat{f}_{dc}$;

Step 6: Operating linear curving fitting of each range cell to further eliminate random estimation error [7].

The flowchart of the proposed algorithm is shown in Fig. 3.5.

3.3.3 Simulation Results and Analysis

In order to demonstrate the validity of the proposed algorithm, both simulation experiments and raw data are processed in this section.

In the simulation experiment, peak amplitude of the azimuth spectrum is set as 100, peak of Target A is 160, and peak of Target B is 40. To better simulate the robustness of the proposed algorithm, white Gaussian noise is added to the spectrum with amplitude of 5.

Figure 3.6a shows the simulated power spectrum of the azimuth data. Target A and B represent two kinds of prominent point targets. From Fig. 3.6b, the samples of Target A and B are far from fitting curve, and the peak of the fitting curve slightly departs from the actual Doppler center due to the impact of exceptional samples. Figure 3.6c shows that exceptional samples are abandoned, and new discrete azimuth spectrum is obtained. Figure 3.6d illustrates that after the second Gaussian curve fitting, the fitting curve is generally the same as the ideal azimuth power spectrum, and the center of the fitting curve is the same as the actual Doppler centroid. Therefore, the Doppler centroid can be estimated by operating traditional energy balancing algorithm. After processing the proposed algorithm, the homogeneity character of the scene is inherently improved, and the estimation precision is increased.

To testify the effectiveness of the proposed algorithm, imaging of real Ku-band airborne SAR data is processed. Figure 3.7a, b show focusing performance before and after proposed algorithm without autofocus technique. It is clear that Fig. 3.7b has a finer resolution, since Doppler centroid is estimated more precisely and RCM is completely corrected. Figure 3.7c offers the optical photo of the scene as a reference. Therefore, the validity of the proposed algorithm is confirmed.

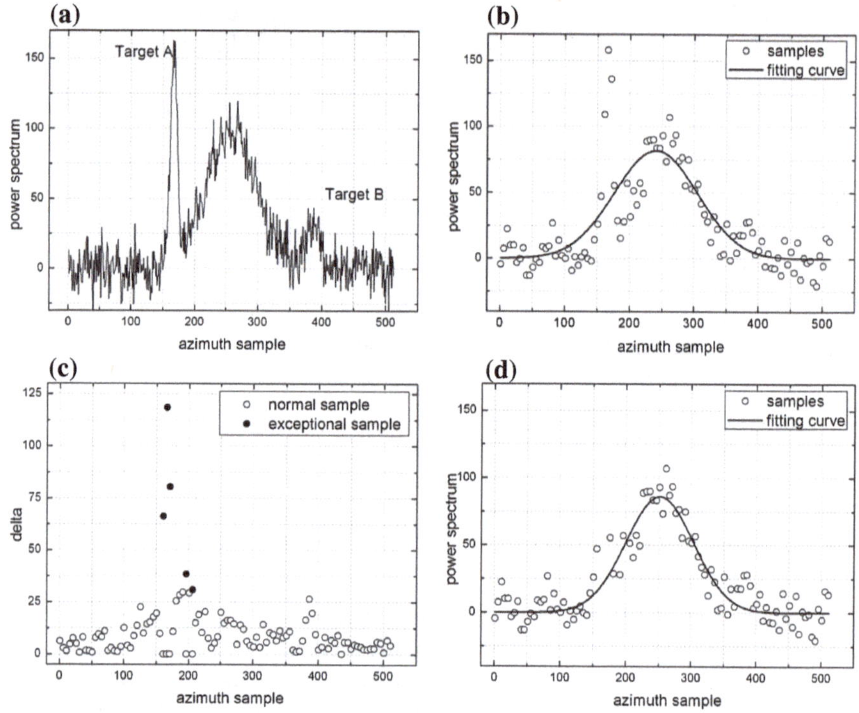

Fig. 3.6 Results of the simulation. **a** Simulated azimuth spectrum. **b** After the first Gaussian curve fitting. **c** Elimination of exceptional samples. **d** After the second Gaussian curve fitting

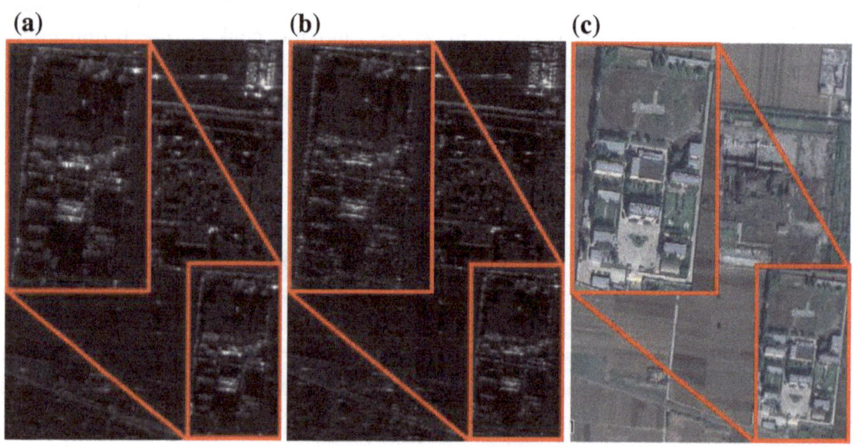

Fig. 3.7 Results of raw data imaging. **a** Traditional algorithm. **b** Proposed algorithm. **c** Optical image

3.4 Multiple Moving Target Indication and Spectrum Extraction

According to the analysis in Sect. 2.3.2, moving targets are classified into three categories by the relationship between its spectrum and the spectrum of the clutter. In real SAR signal processing, the parameters of the moving target is unknown. Therefore, it is rigorous to assume that all three kinds of targets exist in the scene. In order to indicate targets of different kinds, new algorithms must be proposed.

3.4.1 The Two-Step GMTI Algorithm

The principle of the GMTI algorithm for SAR is to suppress the energy of the clutter, and then indicate the moving target according to the result of the amplitude judgment, such as the CFAR detection. Therefore, the main aim of the GMTI algorithm is to suppress the clutter, and raise the SCR of the moving target.

According to Sect. 2.3.2, the moving targets with different motion parameters belong to different types. The targets of Type I are located at the high-band of the PRF, and the clutter are mostly located at the low-band. As a result, the targets of Type I have a high SCR in the range-Doppler domain. On the contrary, targets of Type II and III are submerged by the clutter. The SCR of the targets of Type II and III are low in the range-Doppler domain. After the azimuth compression, the energy of the moving targets of Type II and III are concentrated in the image domain, and the energy of the clutter is distributed in the whole image. If the energy of the moving target is strong, it can be indicated in the image domain. Therefore, a two-step GMTI algorithm is proposed to detect the moving targets of all the three types, and its specific processing steps are shown as follows.

Step 1: Perform the range compression on the echo, and transform it into the range-Doppler domain. In the range-Doppler domain, the spectra of the moving targets of Type I are located at the high-band of the PRF. Extract the signal of the high-band of the PRF using a high-pass filter. If there are multiple trajectories at the high-band of the PRF, each trajectory can be regarded as a moving target.

Perform the azimuth compression on the extracted signal, and detect the moving targets by the CFAR technique. After the detection, the energy of the moving targets should be eliminated from the clutter.

Step 2: After Step 1, the moving targets of Type I have been indicated and extracted. Divide the whole synthetic aperture into two sub-apertures in the range-Doppler domain, and perform the azimuth compression on both sub-aperture data. Two sub-aperture images can be obtained, in which the moving targets have different Doppler histories while the clutter has the same. Therefore, the clutter can be suppressed by the complex subtraction of the two sub-images, and the moving targets can be indicated in the image domain. The targets of Type II and III can be indicated in this step, only if their energy is strong enough to be distinguished after the clutter suppression.

The whole GMTI algorithm is shown above. The reason why the proposed GMTI algorithm uses a two-step indication is that: in real GMTI applications, multiple moving targets with different motion characters must be treated separately. As to the weak targets with fast cross-track velocities (Type I), it is unable to indicate them in the image domain. One the contrary, the image domain algorithm is highly suitable for the indication of the strong targets with slow cross-track velocities and the along-track moving targets.

It is also worth noticing that the proposed algorithm is still unable to indicate the slow and weak moving targets. This is the inherent GMTI limitation of the single-antenna SAR system. In fact, it is still a challenging issue even for the multi-channel SAR system in real applications.

3.4.2 Real Data Processing Results

In this section, the real Ku-band airborne SAR data is used to verify the effectiveness of the proposed strategy. This scene includes roads in both the cross- and along-track directions and a crossroad. Seven moving targets with different motion parameters, which belong to Type I, II, and III, are found in the scene. Moreover, the velocity errors of both the platform and the moving target exist in the data, and they must be compensated during the imaging.

The seven moving targets, labeled from T1 to T7, are illustrated in both the range-Doppler domain and the image domain in Fig. 3.8. The blue lines in Fig. 3.8a indicate the spectrum of the clutter. According to Fig. 3.8a, T1, T2, T3, and T4 move along the cross-track direction. The trajectories of T1 and T2 are located at the high-band of the PRF, and T3 is partially submerged by the clutter in the range-Doppler domain. The trajectory of T4 can be distinguished from the clutter in Fig. 3.8a due to its different slope, whereas it is completely submerged by the clutter. The detection of T4 is the most difficult problem in a single-antenna SAR system. T1 and T2 cannot be seen in Fig. 3.8b since their energy is weaker compared with that of the clutter. T5, T6 and T7 have only along-track velocities and can be found defocused in the azimuth direction in Fig. 3.8b, whereas they cannot be detected in the range-Doppler domain due to that they have no cross-track velocities. We can conclude from Fig. 3.8 that T1 and T2 belong to Type I, T3 belong to Type II, and T4, T5, T6, and T7 belong to Type III. The detection of the moving targets from different types must be conducted differently.

The CFAR detection result of T1 and T2 is shown in Fig. 3.9. The spectrum of the clutter is suppressed by a high-pass filter, and then the rest echo is compressed in the azimuth direction. Particularly, the image of T2 splits into two parts since the spectrum of T2 exceeds the limit of the PRF. It can be noted that moving target of Type I can be successfully indicated by the two-step GMTI algorithm in the range-Doppler domain.

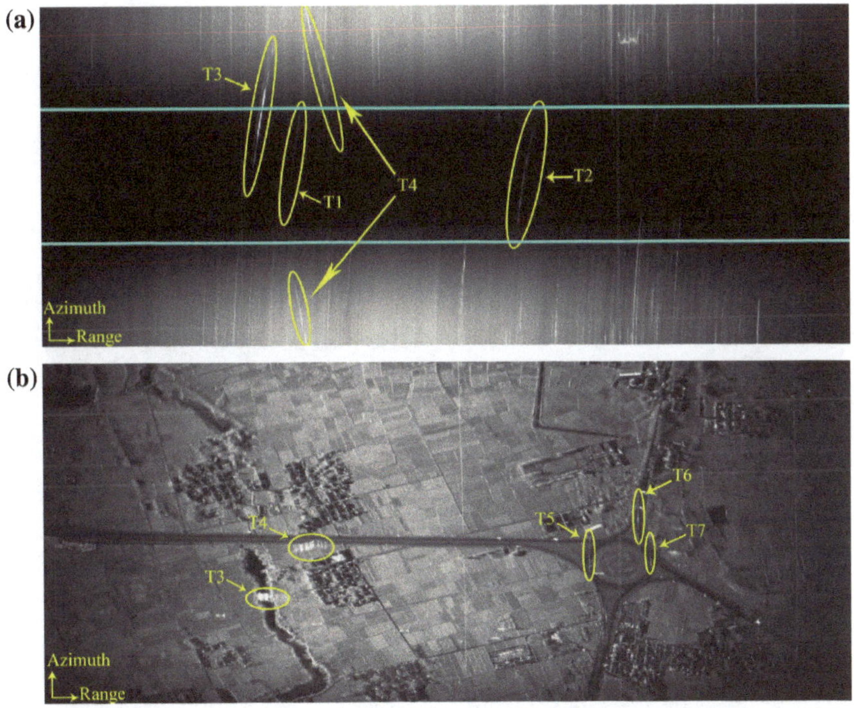

Fig. 3.8 Illustrations of the moving targets in the scene. **a** The moving targets in the range-Doppler domain. **b** The moving targets in the stationary image

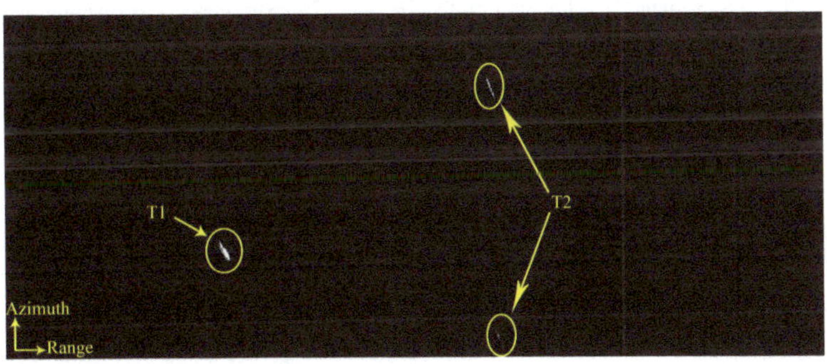

Fig. 3.9 CFAR detection result of T1 and T2

After the detection of T1 and T2, their energy is extracted, and the remaining signal is separated into two sub-apertures. The sub-aperture imaging results are shown in Fig. 3.10. It can be noted from Fig. 3.10 that the locations of the moving targets are different in the two sub-images. The moving targets can be detected by

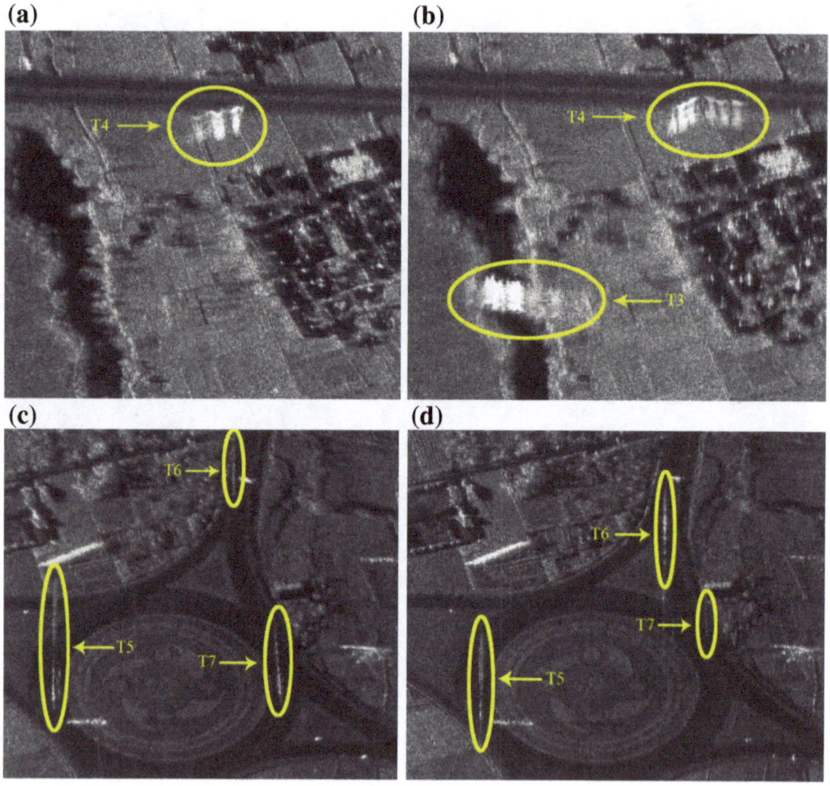

Fig. 3.10 Results of the sub-aperture imaging. **a** Imaging of T3 and T4 in Sub-image I. **b** Imaging of T3 and T4 in Sub-image II. **c** Imaging of T5, T6, and T7 in Sub-image I. **d** Imaging of T5, T6, and T7 in Sub-image II

the complex image subtraction. T3 is not found in Fig. 3.10a because its energy locates only in half of the PRF, as shown in Fig. 3.8a.

The result of the moving target spectrum extraction is shown in Fig. 3.11. T3 and T4 are used as an example to show the effectiveness of the proposed algorithm because the extraction of targets that are submerged by the clutter is very difficult. It can be noted from Fig. 3.11a that the trajectories of T3 and T4 are extracted by the proposed algorithm. Although the energy of the clutter is not completely eliminated, the SCR after the extraction is clearly improved. The imaging of the clutter after the extraction is shown in Fig. 3.11b. The smeared images of the moving targets are disappeared, and the stationary area that masked by the moving targets is reverted.

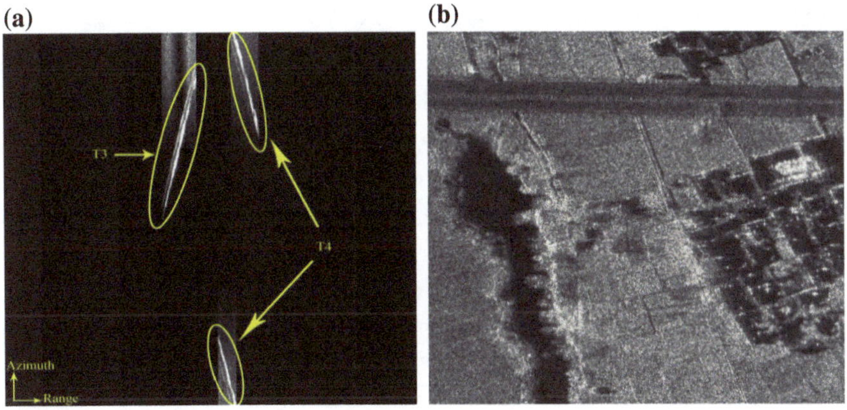

Fig. 3.11 Results of the moving target extraction. **a** The extracted trajectory of T3 and T4 in the range-Doppler domain. **b** Partial imaging of the scene without moving targets T3 and T4

3.5 Fast Moving Target Imaging Algorithm Based on Hough Transform and Third-Order PFT

3.5.1 Principle of the Proposed Algorithm

In conventional GMTIm algorithms [8–10], the basic processing steps: firstly, the RCMC is proceeded to compress the energy of the target into a single range gate; secondly, the parameters of the moving target is estimated by using the time-frequency analysis algorithms. Keystone transform is able to correct the range walk without prior knowledge of the Doppler centroid. However, in the case of the fast moving target, Keystone transform is not able to correct the range walk without the Doppler ambiguity number. Therefore, in the case of the fast moving target imaging, the Doppler ambiguity must be taken into consideration.

In this section, a novel algorithm is proposed to focus the moving targets with fast cross-track velocities. After range compression and range curve correction, we utilize Hough transform to estimate the slope of the range walk trajectory of the moving target. With Hough transform, the additional range walk can be accurately corrected regardless of the Doppler ambiguity, and the cross-track velocity can be estimated. Then, we adopt PFT to estimate the second- and third-order Doppler parameters of the moving target. The imaging resolution of the moving target is improved after the third-order phase error compensation, and the motion parameters, including the cross-track acceleration, can be accurately estimated.

(a) Principle of the Hough transform

Hough transform was proposed by P.V.C. Hough in 1962 [11] for image characteristic detection. It is widely used in digital image processing, especially in dealing

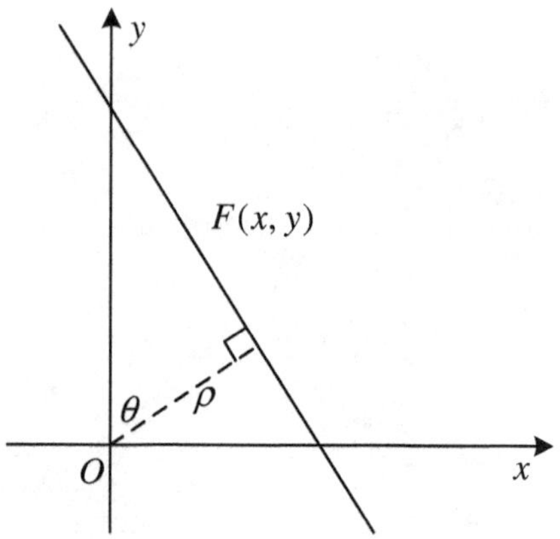

Fig. 3.12 Hough Transform

with straight line indication and parameter estimation. The principle of Hough transform is shown in Fig. 3.12.

In Fig. 3.12, $F(x, y)$ is a function of straight line on Plane (x, y). O is the origin of Plane (x, y). ρ is the length of the normal of $F(x, y)$ to point O, and θ is the angle between normal and axis y. Hough transform is a transform from Plane (x, y) to Plane (ρ, θ). It is supposed that

$$\rho = x \cos\theta + y \sin\theta \tag{3.5.1}$$

By using Eq. 3.5.1, each point in Plane (x, y) corresponds to a sinusoid in Plane (ρ, θ). Therefore, all points in Plane (x, y) correspond to a series of sinusoid interweaved together in Plane (ρ, θ). If points in Plane (x, y) follows the straight line function $F(x, y)$, it appears a peak in Plane (ρ, θ). By indicating the coordinates of the peak, the parameters of $F(x, y)$ can be estimated.

After range compression and range cell curve correction, the echo of moving target is a slant line, and its slope represents linear RCM. By using the Hough transform, the cross-track velocity can be estimated.

(b) Principle of the PFT

Time-frequency analysis is a useful tool in ground moving target parameter estimation, and it is widely adopted in many research articles. Time-frequency analysis can achieve high precision estimation of Doppler parameters of LFM signals, while it will cost a lot of calculation and limit its usage in real-time GMTI project at the same time.

WVD, STFT, FrFT and PFT are all time-frequency analysis algorithms. WVD is the most commonly used algorithm, but it suffers from the interference of cross-terms since it is bilinear transform. FrFT is free cross-term because it is a linear transform, but its estimation precision depends on the resolution of angle division. Both algorithms above are only able to estimate first- and second- order phase, which is not suitable for fast moving target imaging.

PFT is also a linear transform and shares the advantages of FrFT. Besides, PFT is capable of estimating higher order phase parameters of moving targets, which makes it perfect for fast moving target imaging [12]. Suppose $x(n)$ is a discrete signal, its standard DFT is expressed as

$$X(\omega) = \sum_{n} x(n) \exp(-j\omega n) \tag{3.5.2}$$

If $x(n)$ is a monochromatic signal, energy of $x(n)$ is congregated in frequency domain. If $x(n)$ has higher order phase terms, its energy will spread in frequency domain after DFT.

Consider a kth-order PPS $x(n)$, let its expression be

$$x(n) = \exp\left(j\pi \sum_{m=1}^{k} a_m n^m\right) \tag{3.5.3}$$

where a are the coefficients of each order. PFT is defined as

$$X(\alpha_1, \alpha_2, \alpha_3, \ldots, \alpha_k) = \sum_{n} x(n) \exp\left(-\mathrm{j}\pi \sum_{m=1}^{k} \alpha_m n^m\right) \tag{3.5.4}$$

According to the definition of (3.5.4), PFT is an expansion of DFT into kth-order. If coefficients α satisfy condition

$$\alpha_i - a_i (i - 1, 2, 3, \ldots, k) \tag{3.5.5}$$

Energy of PPS $x(n)$ will congregated into a peak in k-order coefficient plane through PFT. By searching the value of α, phase coefficient a can be estimated as

$$(\hat{a}_1, \hat{a}_2, \ldots, \hat{a}_k) = \arg \max_{(\alpha_1, \alpha_2, \alpha_3, \ldots, \alpha_k)} |X(\alpha_1, \alpha_2, \alpha_3, \ldots, \alpha_k)| \tag{3.5.6}$$

The remaining problem of PFT is that higher order estimation introduces more calculation requirements. However, since the Doppler centroid has already been estimated by Hough transform, only the second- and third-order Doppler parameters are estimated in this step.

3.5.2 Processing Steps of the Algorithm

After range compression, the RCM trajectory of the moving target smears among several range cells. In order to obtain a focused image of the moving target, the RCM must be accurately corrected.

The RCM of the moving target can be divided into two parts: the range curve and the additional range walk.

It has been validated in [13] that the platform velocity is the main factor in determining the range curve since that V_y is far smaller than V_a. Therefore, compensating for the range curve caused by the platform is sufficient to the requirement of moving target imaging. The range curve correction filter in range-frequency domain can be expressed as

$$H_{rcmc} = \exp\left(\mathrm{j}\frac{2\pi V_a^2}{cR_0} f_r t_a^2\right) \tag{3.5.7}$$

After range curve correction, the range walk trajectory of the moving target is a straight line. The slope of the range walk trajectory increases with the cross-track velocity, and it is independent of the Doppler ambiguity. Therefore, the actual value of the cross-track velocity can be obtained from the slope even in the presence of the Doppler ambiguity.

Hough transform is an effective tool to detect rectilinear figures, and it is capable of estimating the parameters of the rectilinear figure. During T_a, the cross-track and along-track motions of the moving target in 2-D image domain can be expressed, respectively, as

$$\Delta x = V_r T_a / \frac{c}{2f_s} \tag{3.5.8a}$$

$$\Delta y = PRF \cdot T_a \tag{3.5.8b}$$

Therefore, the tangent of the slope of the range walk trajectory can be expressed as

$$\tan\theta = \frac{\Delta x}{\Delta y} = \frac{V_r \cdot 2f_s}{PRF \cdot \mathrm{c}} \tag{3.5.9}$$

where θ represents the angle between the trajectory and azimuth direction. With the estimation of θ, the cross-track velocity can be estimated as

$$\hat{V}_r = \frac{\tan\hat{\theta} \cdot PRF \cdot \mathrm{c}}{2f_s} \tag{3.5.10}$$

With $\hat{V}_r$, the range walk correction filter in range-frequency domain can be expressed as

$$H_{rwmc} = \exp\left(-\mathrm{j}\frac{4\pi\hat{V}_r}{\mathrm{c}} f_r t_a\right) \tag{3.5.11}$$

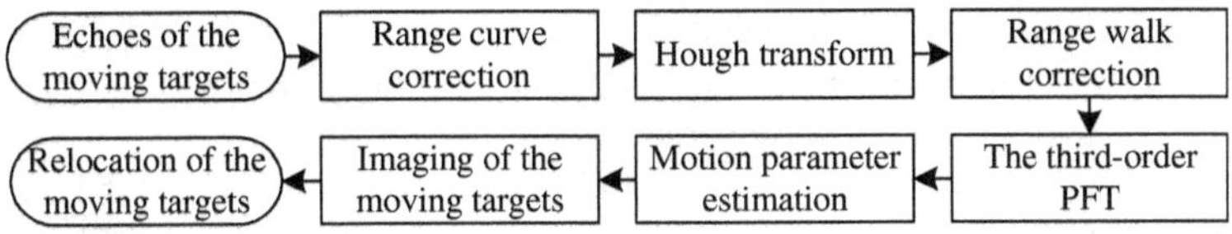

Fig. 3.13 Flowchart of the proposed GMTIm algorithm

After RCMC, the energy of the moving target is concentrated into a single range cell. In order to estimate the third-order Doppler parameter of the moving target, we perform the third-order PFT on the echoes of the moving target.

By using PFT, the second- and third-order phase error can be accurately compensated, and V_y and a_r can be retrieved from the estimations of PFT, respectively, as

$$\hat{V}_y = V_a - \sqrt{-\frac{R_0^2 \lambda \hat{a}_3}{2\hat{V}_r}} \tag{3.5.12a}$$

$$\hat{a}_r = \frac{\lambda (R_0 \hat{a}_3 - \hat{V}_r \hat{a}_2)}{2\hat{V}_r} \tag{3.5.12b}$$

where $\hat{\alpha}_2$ and $\hat{\alpha}_3$ are the second- and third-order Doppler parameter estimations, respectively.

The flowchart of the proposed algorithm is shown in Fig. 3.13.

3.5.3 Results of the Simulation and Real Data Processing

In this section, we simulate a single-antenna SAR system (see Table 3.1) to validate the proposed algorithm. One stationary target, labeled by T0, and three moving targets, labeled by T1, T2, and T3, are set in the scene. The motion parameters of the moving targets are listed in Table 3.3. According to Table 3.3, T1 and T3 represent the moving targets with slow and fast cross-track velocities, respectively,

Table 3.3 Target parameters of the simulation

Target	Motion parameters				
	Nearest slant range (m)	Azimuth location shift (m)	Cross-track velocity (m/s)	Along-track velocity (m/s)	Cross-track acceleration (m/s^2)
T0	1100	10	0	0	0
T1	800	0	10	10	0
T2	900	10	10	0	0
T3	1000	0	50	10	1

and T2 represents the moving target that is not located at the center of the scene. To better testify the improvement of the proposed algorithm, the conventional algorithm in [13] is operated as a comparison.

Figure 3.14a, b show the processing results of the conventional and the proposed algorithm, respectively. For the slow moving target T1, it is focused and relocated

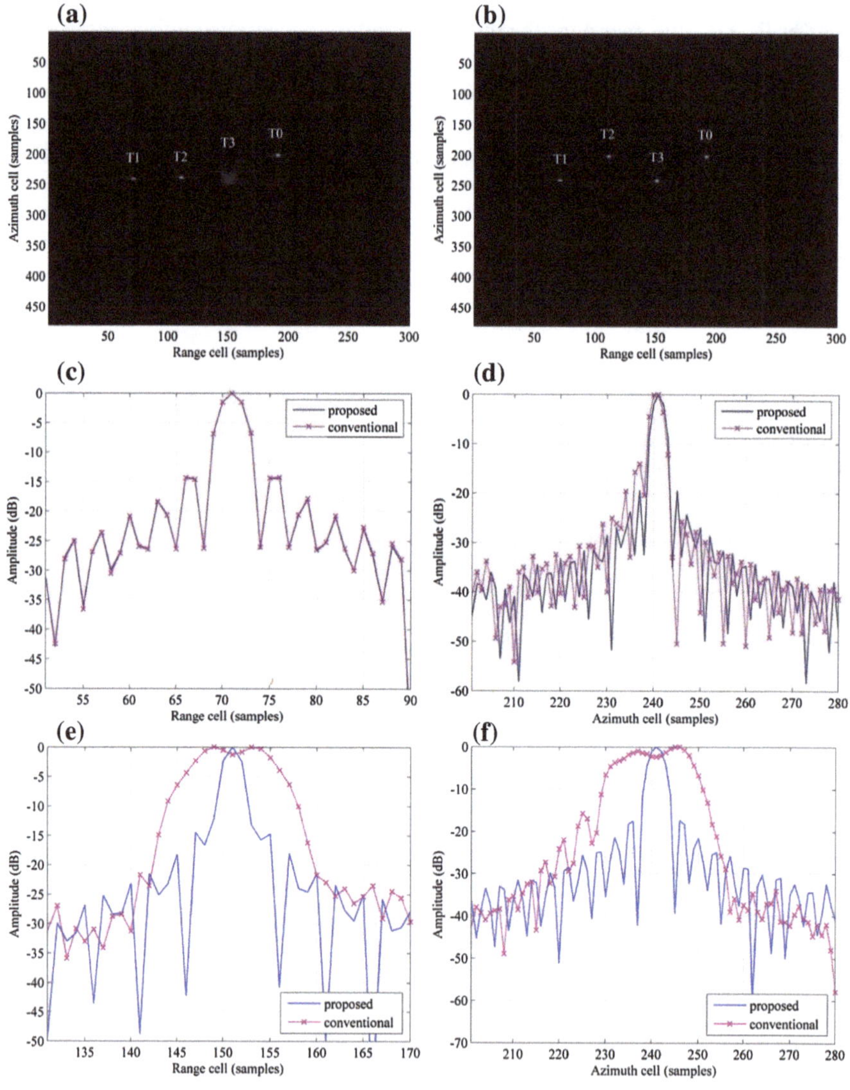

Fig. 3.14 Simulation results of multiple moving target imaging. **a** Processing result of the conventional algorithm. **b** Processing result of the proposed algorithm. **c** Range compression result comparison of T1. **d** Azimuth compression result comparison of T1. **e** Range compression result comparison of T3. **f** Azimuth compression result comparison of T3

in both Fig. 3.14a, b. However, the image of T3 in Fig. 3.14a is severely smeared since the range walk is not completely corrected by Keystone transform and the third order phase error is not compensated. On the contrary, T3 is well-focused in Fig. 3.14b. From the comparison, we can conclude that the proposed algorithm is effective in focusing both the slow and fast moving targets. The relocation of T2 in Fig. 3.14a is inaccurate since the energy balance filter method cannot provide the accurate location if the target is not located at the center of the azimuth aperture. However, this problem is avoided by Hough transform, as shown in Fig. 3.14b.

Figure 3.14c, d further compare the range and azimuth compression results of T1. The side-lobe asymmetry is not negligible in Fig. 3.14d although V_r of T1 is slow. The range and azimuth compression results of T3 are compared in Fig. 3.14e, f. It can be noted that the range and azimuth resolutions of T3 by the conventional algorithm are both deteriorated by the range walk and the third-order phase error.

Motion parameter estimations are listed in Table 3.4. The estimations of V_r and f_{dc} of T2 are incorrect by the conventional algorithm since the azimuth location affects the accuracy of estimation. The estimations of V_r and f_{dc} of T3 by the conventional algorithm depart from the actual values because of the Doppler ambiguity, and the estimation of V_y of T3 by the conventional algorithm is incorrect since the moving target is assumed with constant velocities in [13]. On the contrary, the proposed algorithm provides accurate motion parameter estimations of all the targets.

Table 3.4 Comparison of the motion parameter estimations

Target		Parameters			
		Doppler centroid (Hz)	Cross-track velocity (m/s)	Along-track velocity (m/s)	Cross-track acceleration (m/s^2)
T1	Actual value	133.33	10	10	0
	Traditional algorithm	133.00	9.9750	10.0333	0
	Proposed algorithm	133.30	9.9975	10.0667	0
T2	Actual value	133.33	10	0	0
	Traditional algorithm	119.1667	8.9375	0	0
	Proposed algorithm	133.30	9.9975	0	0
T3	Actual value	666.67	50	10	1
	Traditional algorithm	266.00	19.9500	15.5703	0
	Proposed algorithm	666.20	49.9650	9.7261	1.0210

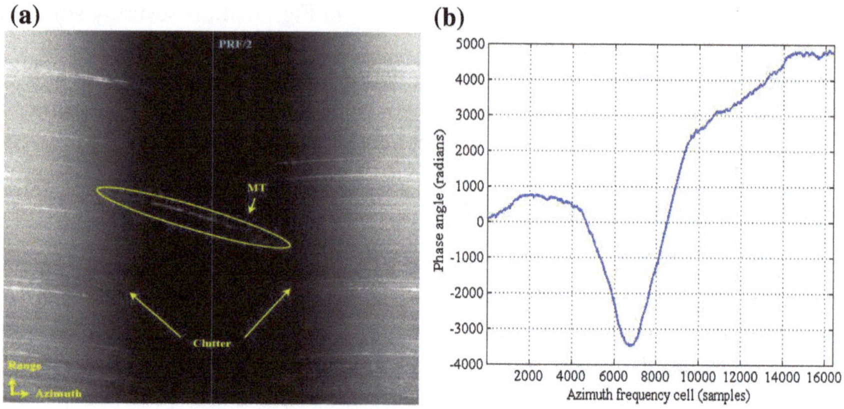

Fig. 3.15 Spectrum analysis of moving target in real data. **a** Doppler spectrum of the data after range compression. **b** Azimuth phase angle of the moving target

To further verify our algorithm, we use real Ku-band airborne SAR/GMTI data in Sect. 3.4.2 to further demonstrate the robustness of the proposed algorithm. The selected data contains the echo of a vehicle T2 moving along a highway in the scene. T2 is not a controlled target so that the parameter estimation accuracy cannot be compared in this section.

Figure 3.15a shows the Doppler spectrum of the selected data after range compression. From Fig. 3.15a, we can see that the trajectory of moving target, which is tagged by an ellipse, locates at the high-band of PRF due to the Doppler centroid shift, and the Doppler spectrum of the moving target exceeds the limit of PRF. Figure 3.15b shows the unwrapped azimuth phase angle of the target. The azimuth phase angle of the moving target is not a standard quadratic curve, i.e., there are higher order phase errors. Therefore, the necessity to compensate for the third-order phase error in real data imaging is confirmed.

Figure 3.16a shows the stationary image of the scene without the processing of the moving target. As tagged by the ellipses, the moving target is smeared and dislocated from the highway. Moreover, the image of the moving target splits into two parts because its spectrum expands neighboring PRF. Figure 3.16b shows that with the proposed algorithm, the moving target is focused and relocated on the highway. Therefore, the effectiveness of the proposed algorithm in real airborne SAR data is demonstrated.

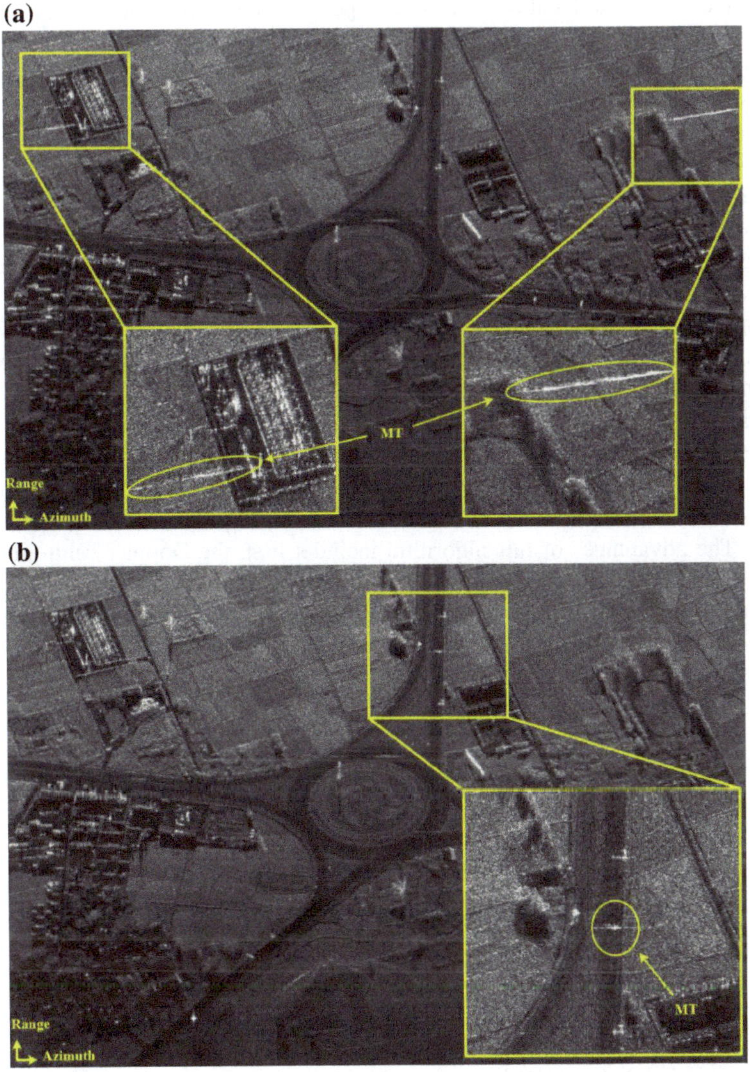

Fig. 3.16 Imaging results of the data without and with the proposed algorithm. **a** Stationary image of the scene without moving target imaging. **b** Imaging result of the moving target by the proposed algorithm

3.6 Conclusion

The fast moving target indication and imaging have been discussed in stripmap SAR. Firstly the echo model of a moving target with fast cross-track velocity has been established, and the Doppler centroid ambiguity problem has been analyzed. It is concluded that the location of azimuth spectra cannot reflect the actual value of Doppler centroid, while the RWM is not affected by the Doppler centroid ambiguity.

Based on the echo analysis, a Doppler centroid estimation algorithm based on curve fitting, a multiple target indication and extraction method, and a fast moving target imaging algorithm based on Hough transform and third-order PFT have been proposed.

The Doppler centroid estimation algorithm is a modification of the energy balancing algorithm. By eliminating the strong targets in the scene, the homogeneity of the scene is alleviated, and the estimation accuracy is improved.

Consider the multiple target indication with different motion parameters, the indication algorithm uses two steps to detect the targets and extract their spectra. The indication and extraction are the foundation of GMTIm, and is highly suitable for practical applications.

The fast moving target imaging algorithm uses Hough transform to estimate the slope of the RWM to derive the cross-track velocity and actual Doppler centroid, and uses the third-order PFT to estimate the azimuth parameters and refocus the target. The advantages of this algorithm include: first, the Doppler centroid ambiguity problem is solved; second, the third-order phase error caused by the fast cross-track velocity is corrected; third, the targets that not located at the center of the scene can be accurately relocated by using PFT; fourth, the cross-track acceleration can be correctly estimated.

The algorithms proposed in this chapter are effective for both the fast moving targets and slow moving targets. Both simulations and real data have been used to demonstrate the effectiveness of these algorithms. However, for the targets with slow velocities and weak RCS, these algorithms show poor performances due to the limitations of single-antenna systems.

References

1. Bamler R (1991) Doppler frequency estimation and the Cramer-Rao bound. IEEE Trans Geosci Remote Sens 25(3):385–390
2. Li FK, Held DN, Curlander JC, et al (1985) Doppler parameters estimation for spaceborne synthetic-aperture radars. IEEE Trans Geosci Remote Sens GE-23(1):47–56
3. Madsen SN (1989) Estimating the Doppler centroid of SAR data. IEEE Trans Aerosp Electronic Syst 25(2):134–140
4. Lin MY (1986) Optimal Doppler centroid estimation for SAR data from a quasi-homogeneous source. IEEE Trans Geosci Remote Sens GE-24(2):1022–1025
5. Long T, Lu Z, Ding Z et al (2011) A DBS Doppler centroid estimation algorithm based on entropy minimization. IEEE Trans Geosci Remote Sens 49(10):3703–3712
6. Wei S, Fang Z, Wang H (2004) An improved method of Doppler centroid estimation in SAR. In: IEEE international geoscience and remote sensing symposium, September 2004, vol 2, pp 1530–1533
7. Sun H, Gu H, Su W et al (2002) Two improved Doppler centroid estimation techniques in imaging through Doppler beam sharpening. ACTA Armamentarii 23(1):49–53

8. Jao JK (2001) Theory of synthetic aperture radar imaging of a moving target. IEEE Trans Geosci Remote Sens 39(9):1984–1992
9. Zhu S, Liao G, Qu Y et al (2011) Ground moving targets imaging algorithm for synthetic aperture radar. IEEE Trans Geosci Remote Sens 49(1):462–477
10. Xu R, Zhang D, Hu D et al (2012) A novel motion parameter estimation algorithm of fast moving targets via single-antenna airborne SAR system. IEEE Geosci Remote Sens Lett 9 (5):920–924
11. Hough PVC (1962) Method and means for recognizing complex patterns. US Patent 3069654, Dec 1962
12. Djurovic I, Thayaparan T, Stankovic L (2008) SAR imaging of moving targets using polynomial Fourier transform. IET Signal Proc 2(3):237–246
13. Yang J, Huang X, Jin T et al (2011) New approach for SAR imaging of ground moving targets based on a Keystone transform. IEEE Geosci Remote Sens Lett 8(4):829–833

Chapter 4
Ground Moving Target Indication and Imaging in WAS Mode

Abstract This chapter mainly discusses the GMTI and GMTIm algorithms in WAS mode. The signal mode of moving target is established in WAS mode, and a new system design scheme is presented to maintain the sharpening ratio constant without increasing the system and signal processing complexity. Moreover, a real-time DBS algorithm is proposed based on CZT, and a GMTI and motion parameter estimation algorithm based on multiple revisits has been presented. Finally, simulations and real data have been utilized to prove the effectiveness of these algorithms.

4.1 Introduction

Besides the stripmap SAR mode, scan mode is also a typical working mode of a SAR system. Different from the fixed antenna LOS in the stripmap SAR mode, the LOS scans with a rotating speed along the azimuth direction in the scan SAR mode. Therefore, larger observation areas can be obtained in the scan SAR mode under the same observation time. Also, the reciprocation of the antenna allows the system to revisit the same target, and more information of the target can be acquired to the target indication and imaging. Based on these two advantages, scan SAR mode is more suitable for moving target processing compared with the stripmap SAR mode, and it is also called WAS mode.

WAS mode can be classified into phased-array scanning mode and mechanic scanning mode. In the phase-array scanning mode, the LOS is controlled by altering the phase of the transmitted signal in each channel, and multi-channel signal receive can be realized [1]. However, the phase-array antenna is relatively expensive, and the signal processing burden is very heavy. The mechanic scanning mode achieves azimuth scanning by mechanic rotation of the single antenna. The antenna control accuracy of the mechanic scanning mode is worse than the phased array scanning mode, but it has a simpler system structure, and can be achieved using the same stripmap SAR system. Therefore, this chapter mainly focused on the mechanic scanning mode.

J. Yang, *Study on Ground Moving Target Indication and Imaging Technique of Airborne SAR*, Springer Theses, DOI 10.1007/978-981-10-3075-8_4

In the mechanic scanning mode, the traditional RD and CS algorithms of the stripmap SAR mode cannot be used to acquire a focused SAR image. DBS algorithm is used to improve the azimuth resolution. This chapter firstly analyzed the geometry and signal model in the mechanic scanning mode, and the principle and processing steps of the DBS algorithm are introduced. To maintain the sharpening ratio constant, a system design scheme suitable for the airborne SAR mechanic scanning mode is presented, and a real-time DBS imaging algorithm based on CZT is proposed. Since the GMTI processing in the mechanic scanning mode is operated in the imagery domain, the qualities of the images are highly important for the indication.

In the phased-array scanning mode, the indication and parameter estimation of a moving target is achieved by using multi-channel GMTI algorithms, such as DPCA. However, these algorithms are not suitable for the mechanic scanning mode since only single-channel images can be obtained. To solve this problem, a GMTI and parameter estimation algorithm based on multiple revisits is proposed in this chapter. The indication of the moving targets in each image is achieved by the Spectrum filtering algorithm, and the motion parameters are estimated by using the location changes of the targets between each image. Real SAR data is used to prove the effectiveness of this algorithm.

This chapter is organized as follows. In Sect. 4.2, the signal model and the system design scheme are discussed. Section 4.3 introduces the proposed algorithm based on CZT. In Sect. 4.4, the GMTI and parameter estimation algorithm based on multiple revisits is presented. Finally, conclusive remarks are provided in Sect. 4.5.

4.2 Key Technique in Airborne SAR Mechanic Scanning Mode

4.2.1 Mechanic Scanning Mode and the Principle of DBS

In the mechanic scanning mode, in the existence of the antenna scanning, a target in the observation area has a shorter observation time compared with the stripmap SAR mode. Therefore, the resolution of the mechanic scanning mode is lower than that of the stripmap SAR mode, and the imaging geometry is different [2].

The geometry of mechanic scanning system is illustrated in Fig. 4.1. The aircraft flies at velocity V_a along axis x. The angle between beam center and axis y is θ. The 3-dB azimuth beam-width is denoted as $\Delta\theta$. The antenna is scanning along azimuth direction at constant speed $\hat{\theta}$. The nearest slant range from O to the ground is R_0, and the instantaneous slant range from O to B is denoted as $R(t_a)$.

The Doppler centroid along the LOS can be expressed as

$$f_{dc}(\theta) = \frac{2V_a \sin\theta}{\lambda} \tag{4.2.1}$$

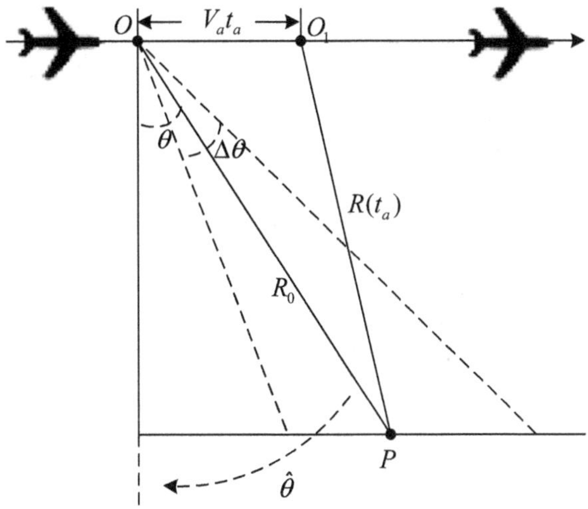

Fig. 4.1 Geometry of a mechanic scanning system

The Doppler bandwidth $\Delta f_d(\theta)$ can be expressed as

$$\Delta f_d(\theta) = f_{dc}\left(\theta + \frac{\Delta\theta}{2}\right) - f_{dc}\left(\theta - \frac{\Delta\theta}{2}\right) = \frac{2V_a \cos\theta}{\lambda}\Delta\theta \tag{4.2.2}$$

According to Eq. 4.2.2, the Doppler bandwidth is linearly correlated to the beam-width. The principle of DBS technique is: since targets from different azimuth cells have different Doppler frequency, the division of azimuth spectrum can be regarded as the division of azimuth beam. Therefore, the azimuth angle resolution can be improved by operating narrowband filtering of azimuth frequency band. In practice, narrowband filter can be achieved by Fourier transform.

An important parameter in DBS imaging is Sharpening ratio. Sharpening ratio N represents the ratio of bandwidth to sub-bandwidth, which determines the azimuth resolution. Sharpening ratio N can be defined as

$$N = \frac{\Delta f_d}{\delta f_d} = \frac{\Delta\theta}{\delta\theta} \tag{4.2.3}$$

where δf_d and $\delta\theta$ denote Doppler resolution and beam resolution, respectively. δf_d can be defined as

$$\delta f_d = \frac{1}{T_a} = \frac{M}{PRF} \tag{4.2.4}$$

where T_a is azimuth synthetic time, M is the number of coherent pulses [3]. Substituting Eq. 4.2.2 and Eq. 4.2.4 into Eq. 4.2.3, it yields

$$N = \frac{2V_a \cos\theta \cdot \Delta\theta \cdot M}{\lambda \cdot PRF} \tag{4.2.5}$$

In practice, the narrowband filter of a traditional DBS algorithm can be achieved by Fourier transform. Therefore, the DBS algorithm is highly simple and suited for real-time processing. The processing steps of the traditional DBS algorithm are illustrated in Fig. 4.2. After imaging each frame by DBS, operations such as non-coherent integration, coordinate transformation, and image stitching are also required [4]. However, these steps are not relevant to the imaging quality of each frame, so these are not discussed in this chapter.

In Fig. 4.1, $R(t_a)$ can be expressed as

$$R(t_a) = \sqrt{R_0^2 + (V_a t_a)^2 - 2R_0 V_a t_a \sin\theta} \tag{4.2.6}$$

After Taylor expansion, $R(t_a)$ can be approximated as

$$R(t_a) = \sqrt{R_0^2 + (V_a t_a)^2 - 2R_0 V_a t_a \sin\theta} \approx R_0 - V_a t_a \sin\theta + \Phi\left(t_a^2\right) \tag{4.2.7}$$

where $\Phi\left(t_a^2\right)$ is the quadratic azimuth phase and can be neglected in mechanic scanning system.

The echo of a point target in mechanic scanning system can be expressed as

$$s_0(t_a, t_r) = P_0 \exp\left(-j4\pi \frac{R(t_a)}{\lambda}\right) \exp\left(j\pi K_r \left(t_r - \frac{2R(t_a)}{c}\right)^2\right) \tag{4.2.8}$$

where P_0 represents backscattering coefficient, λ denotes wavelength, K_r stands for frequency modulation rate, c is light speed, t_a and t_r denote azimuth and range time, respectively.

Substitute Eq. 4.2.7 into Eq. 4.2.8, and transform Eq. 4.2.8 into range-frequency domain, then it can be rewritten as

$$\begin{aligned} s_0(t_a, f_r) = {} & P_0 \exp\left(j\frac{\pi}{K_r} f_r^2\right) \exp\left(j\frac{4\pi K_r}{c^2}\left(R_0^2 + V_a^2 t_a^2 + 2R_0 V_a t_a\right)\right) \\ & \cdot \exp\left(-j\frac{4\pi}{c}(f_r + f_c)R_0\right) \exp\left(j\frac{4\pi}{c}(f_r + f_c)V_a \sin\theta t_a\right) \end{aligned} \tag{4.2.9}$$

The last term of Eq. 4.2.9 is the RCM, and must be corrected during DBS imaging.

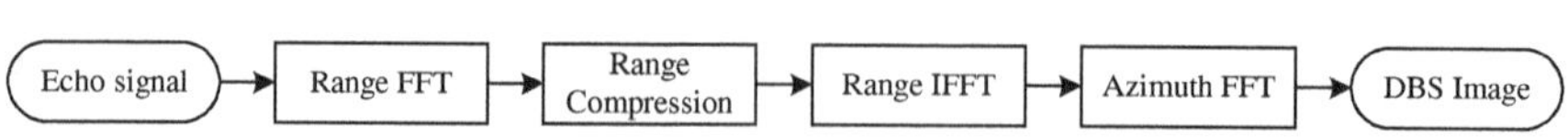

Fig. 4.2 Flowchart of the traditional DBS algorithm

4.2.2 System Design Scheme of the Airborne Mechanic Scanning SAR

An important issue of the mechanic scanning mode is to maintaining N constant. Equation 4.2.5 is the expression of N. In an airborne SAR system, V_a, $\Delta\theta$ and λ are constant values. If $\frac{M}{PRF}$ is also constant, it can be noted that the sharpening ratio changes with the value of θ. And it is not acceptable since the resolution of one image differs at different scanning angles.

There are mainly two methods to solve this problem. First, set the pulse integration amount M constant, and let the PRF changes with the scanning angle [5]; second, set the PRF constant, and M changes with the scanning angle [6]. The first method increases the complexity of the radar system, while the second method increases the signal processing burden by using DFT. Algorithms are also proposed to adjust the sharpening ratio with both PRF and M constant, such as pre-filtering algorithm [7] and frequency-domain interpolation algorithm [8]. However, both algorithms increase the complexity of the DBS algorithm.

The mechanic scanning mode is originally applied on the fire-control radar, of which the typical antenna scanning range is 30°–75°. Under this range, the value of $\cos\theta$ varies from 0.259 to 0.866. In this condition, the sharpening ratio is significantly changed.

In the application of the mechanic scanning mode into airborne SAR, we wish to maintain the sharpening rate constant without increasing the complexity of both the system structure and the signal processing by the optimizing the system design.

According to article [2, 9], a new system design scheme is presented. In this scheme, the scanning range is −18° to 18°. In this condition, the value of $\cos\theta$ varies from 0.951 to 1. Since that DBS is not a high resolution imaging algorithm, the sharpening ratio under this scheme can be viewed as constant. The system parameters are illustrated in Table 4.1.

According to the parameters in Table 4.1, the radar works at Ku-band, and the central range is 15,455 m. As to a target in the observation area, the echo pulse amount is $N_t = \frac{\Delta\theta}{\dot{\theta}} PRF = 388.8 \approx 389$. Consider the impact of coherent sparkle noise, the target is not integrated during the full synthetic aperture in general cases. In the coherent pulse amount is small, a target can be indicated in multiple coherent

Table 4.1 System parameters of airborne SAR mechanic scanning mode

System parameters	Value	System parameters	Value
PRF	2000 Hz	Platform height	4000 m
Downwards angle	15°	Platform velocity	110 m/s
Scanning range	−18°–18°	Wavelength	0.0195 m
Beam-width	3.5°	Time-width	40 us
Scanning speed	18°/s	Range sampling rate	100 MHz

pulse interval (CPI), which not only increases the indication possibility but also eliminates the coherent sparkle noise. Since the coherent integration processing is operated by FFT, set $N_t = 384$, then can be valued as 128.

As to different scanning angle, the Doppler bandwidth is 632.4–664.95 Hz. Substitute the values of Δf_d and M into Eq. 4.2.5, the value range of the sharpening ratio is 40.47–42.56 and the azimuth resolution varies from 22.183 to 23.382 m. Therefore, it is proved that the sharpening ratio is almost constant.

4.3 Real-Time DBS Algorithm Based on CZT

4.3.1 The Modification of Traditional DBS Algorithm

According to Sect. 4.2.1, the image proceeded by DBS algorithm is a false image in the azimuth frequency domain. Therefore, in the WAS mode, the GMTI is operated in the image domain, and DBS algorithm is highly important in the GMTI performance.

In conventional DBS imaging algorithms, RCM is neglected to simplify imaging process since azimuth synthetic time within each azimuth processing window is so short that RCM is negligible. As the increasing demand of azimuth resolution, RCM correction process is induced into DBS algorithm. In these modified algorithms [10], Doppler centroid is estimated in each azimuth processing window. However, Doppler centroid estimation algorithms, such as energy balancing method and entropy minimization method, induce complicate operations, such as data transpose, correlation and iteration, which make DBS imaging not applicable to real-time processing. In addition, if Doppler centroid is not correctly estimated, RCM cannot be perfectly corrected.

Besides the modification of the introduction of RCMC, article [11] also proposed to dechirp the azimuth signal so as to increase the azimuth resolution. However, according to our analysis and simulation, the azimuth second-order phase error is negligible, and the deramping operation is contrary to the efficiency of the DBS algorithm.

4.3.2 Principle of the Proposed Algorithm

In this section, the principle and processing steps of the proposed algorithm are introduced. The aim of the implementation of CZT is to avoid the use of Doppler centroid estimation algorithms. Thus, the processing of DBS algorithm is simplified, and Doppler centroid estimation error is circumvented.

Transform Eq. 4.2.9 into the range-frequency domain, it yields

$$s_r(t_a, f_r) = P_0 \exp\left(-j\frac{4\pi}{\lambda}R_0\right)\exp\left(j\pi\frac{f_r^2}{K_r}\right)\exp\left(-j\frac{4\pi R_0}{c}f_r\right)\exp\left(j\frac{4\pi}{c}V_a \sin\theta(f_c + f_r)t_a\right) \tag{4.3.1}$$

where f_r denotes range frequency. In Eq. 4.3.1, the last exponential term is the RCM in range-frequency domain. It can be noted that the RCM is the correlation of range frequency and azimuth time.

Perform transform of t_a as

$$t_a = \frac{f_c}{f_c + f_r}\tau_a \tag{4.3.2}$$

According to Sect. 2.5, Eq. 4.3.2 is the definition formula of Keystone transform. Keystone transform is capable of adjust the RCM without the estimation of Doppler centroid, which circumstances the Doppler centroid estimation processing steps and the impact of the Doppler centroid estimation error.

However, Keystone transform also faces the problems of Doppler centroid aliasing and interpolation. The Doppler centroid aliasing does not happen if

$$K_v = \frac{PRF}{V_a} > \frac{2\sin\theta}{\lambda} \tag{4.3.3}$$

where K_v represents the PRF-to-velocity ratio. According to the parameters in Table 4.1, K_v can be calculated as 31.69. In a typical airborne SAR, the value of K_v is larger than 32, therefore the Doppler centroid aliasing can be circumstanced.

The interpolation is also an important problem of Keystone transform. The interpolation is not suited for real-time processing, thus it is not suited for the DBS modification. Therefore, a novel DBS algorithm based on CZT is proposed in this chapter [12]. The proposed algorithm can be operated by using complex multiplication and FFT, which is highly suitable for real-time processing.

CZT is an extensive z-transform which non-uniformly samples along a spiral [13]. The CZT of signal $x(n)$ can be defined as

$$X(z) = \sum_{n=0}^{N-1} x(n)z^{-n} \tag{4.3.4}$$

where $z = e^{-j\frac{2\pi}{N}}$, and it denotes the N uniformly sampling on the unit circle. The definition of CZT is

$$z_k = AW^{-k}, k = 0, \ldots, M - 1 \tag{4.3.5}$$

where A_0 is the vector radius length of starting sampling point, θ_0 is the phase of starting sampling point, W_0 is the extensional ratio of spiral, φ_0 represents the angle difference between adjacent sampling points, and M is the complex spectrum samples.

Substitute Eq. 4.3.5 into Eq. 4.3.4, it yields

$$X(z_k) = \sum_{n=0}^{N-1} x(n)\left(AW^{-k}\right)^{-n} \tag{4.3.6}$$

According to the Bluestein equation, Eq. 4.3.6 can be derived as

$$X(z_k) = W^{\frac{k^2}{2}} \sum_{n=0}^{N-1} x(n)A^{-n}W^{\frac{n^2}{2}}W^{-\frac{(k-n)^2}{2}} = W^{\frac{k^2}{2}} \sum_{n=0}^{N-1} g(n)h(k-n) = W^{\frac{k^2}{2}}g(k) * h(k) \tag{4.3.7}$$

where $g(n) = x(n)A^{-n}W^{\frac{n^2}{2}}$, $h(n) = W^{-\frac{n^2}{2}}$. Perform range compression in the range frequency domain, and set the signal as $s_{rc}(t_a, f_r)$. Perform CZT to the azimuth signal in the range-frequency azimuth-time domain, the RCMC can be achieved. Set the signal after CZT as $s_{rc}(\tau_a, f_r)$, the processing steps of the CZT is shown in Fig. 4.3.

RCMC can be realized by CZT rapidly by setting up parameters of CZT as

$$M = N; A_0 = W_0 = 1; \theta_0 = 0; \varphi_0 = \frac{f_c + f_r}{f_c}\frac{2\pi}{M} \tag{4.3.8}$$

Therefore, the proposed algorithm requires simple processing steps, and can be operated by FFT and complex multiplication. Processing steps of CZT is split and combined together with the DBS algorithm, which makes it very suitable for real-time project.

Simulations are performed to prove that RCM can be completely corrected by the proposed algorithm. Simulation parameters are listed in Table 4.1. Nine point targets are simulated, and conventional DBS algorithm [5] is performed as a comparison.

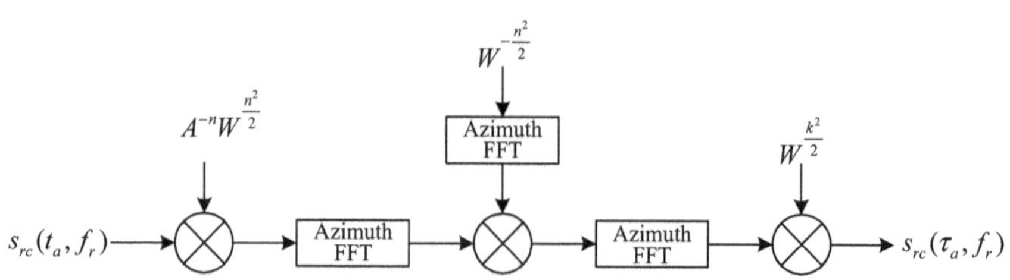

Fig. 4.3 Flowchart of RCMC based on CZT

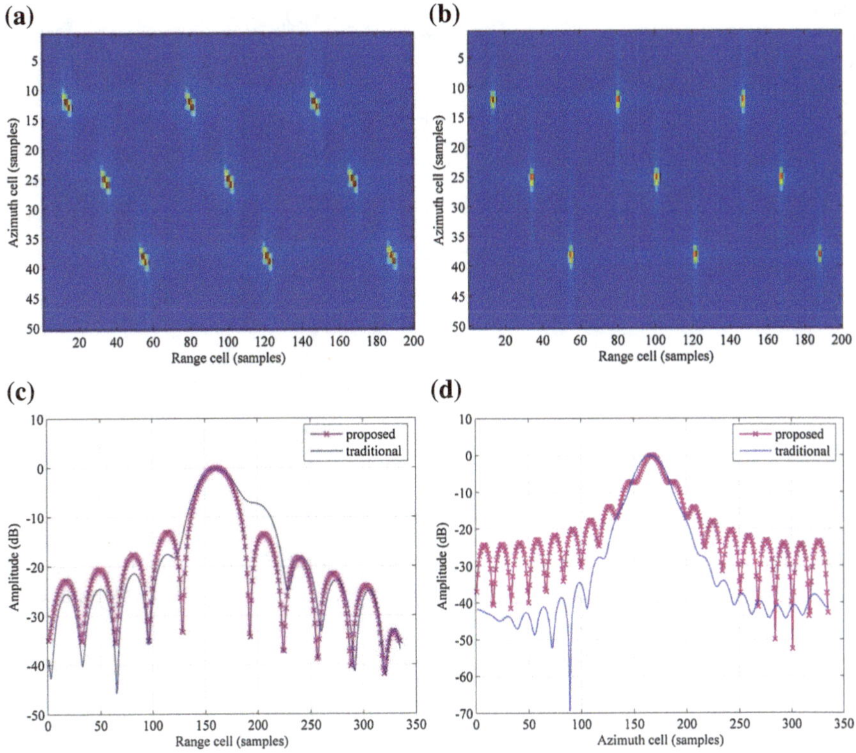

Fig. 4.4 Comparison results of simulations. **a** Conventional algorithm simulation. **b** Proposed algorithm simulation. **c** Range slice of both algorithms. **d** Azimuth slice of both algorithms

Figure 4.4a, b show the simulation results of nine point targets imaging of conventional and proposed algorithms. The squint angle in the simulation is 18°. Targets in Fig. 4.4a are smeared in azimuth since RCM is not corrected. According to Fig. 4.4b, RCM are completely corrected by the proposed algorithm, and the imaging resolution is significantly improved. Figure 4.4c, d further illustrate the compression performances of both algorithms. Thus, the imaging performance of the proposed algorithm is proved.

4.4 GMTI and Parameter Estimation Algorithm Based on Multiple Revisits

The along-track velocity of a moving target mainly causes the azimuth second-order phase error, and the impact can be neglected in DBS imaging. Therefore, the cross-track velocity is analyzed in the GMTI of mechanic scanning mode. The geometry of a moving target in mechanic scanning mode is shown in Fig. 4.5.

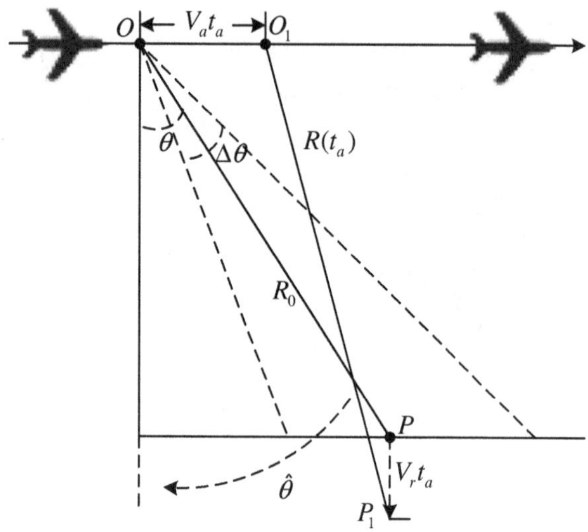

Fig. 4.5 Geometry of a moving target in the mechanic scanning mode

In Fig. 4.5, the cross-track velocity of moving target P is V_r. After azimuth time t_a, the platform flies from O to O_1, while the target moves to P_1. The instantaneous slant range $R(t_a)$ can be expressed as

$$R(t_a) = \sqrt{(R_0 \cos\theta + V_r t_a)^2 + (R_0 \sin\theta - V_a t_a)^2} \tag{4.4.1}$$

After Taylor expansion, $R(t_a)$ can be approximated as

$$R(t_a) \approx R_0 - V_a t_a \sin\theta + V_r t_a \cos\theta \tag{4.4.2}$$

Substitute Eq. 4.4.2 into Eq. 4.2.8, the Doppler centroid of a moving target can be expressed as

$$f_{dc}(\theta) = \frac{2V_a \sin\theta}{\lambda} - \frac{2V_r \cos\theta}{\lambda} \tag{4.4.3}$$

Compared with Eq. 4.2.1, in the existence of V_r, the Doppler centroid of a moving target is shifted from that of a stationary target. The shift will result in the dis-location of a moving target in the azimuth direction in the DBS image. Therefore, the moving target can be indicated by using the Doppler filtering algorithm. The specific processing steps are as follows.

Step 1: Perform DBS imaging of the raw data to obtain the false image;

Step 2: Estimate the Doppler centroid of the clutter in the azimuth-frequency domain, and derive the LOS of the antenna;

Step 3: Eliminate the spectrum of the clutter by using the band-pass filter;

Step 4: Perform CFAR detection of the image to detect the moving targets.

The four steps above are used in the images of each frame. In order to increase the possibility of detection and decrease the false alarm rate, detection results from each frame can be correlated to further judge the existence of the moving targets.

After the indication of the moving targets, the motion parameters are estimated by using the multiple revisits. The specific processing steps are as follows.

Step 1: Perform GMTI of each DBS image to locate each moving target;

Step 2: Read the scanning angle from the subsidiary data to obtain two DBS images of neighboring visit;

Step 3: Perform image registration of the two images. There are mainly two methods: first, operate cross-correlation of the two images to search for the peak; second, use the feature point match method;

Step 4: Calculate the location change ΔN_r of the target in two images. Suppose the range sampling interval is ρ_r, the time interval of the imaging of the two images is Δt, the cross-track velocity can be estimated as

$$\hat{V}_r = \frac{\Delta N_r \rho_r}{\Delta t} \tag{4.4.4}$$

Step 5: By using the estimated $\hat{V}_r$, the Doppler centroid shift Δf_{dc} can be calculated according to Eq. 4.4.3. Suppose coherent pulses is M, the azimuth dis-location ΔN_a can be calculated as

$$\Delta N_a = \frac{\Delta f_{dc} M}{PRF} \tag{4.4.5}$$

Step 6: Re-locate the target into its actual location. If there are multiple targets, repeat these steps.

4.5 Conclusion

The WAS GMTI mode has the advantages of wide-swath, real-time processing, and multiple revisits. Therefore, it is an effective mode to indicate the ground moving targets in the SAR system. The advantages of mechanic scanning mode include simple system structure, low cost and single channel. The main topic of this chapter is focused on the mechanic scanning mode.

First of all, the working scheme of the mechanic scanning mode and the principle of the DBS imaging algorithm have been introduced. To obtain the constant sharpening ratio, a new system design scheme has been presented, which does not increase the complexities of the system and the signal processing.

Secondly, a novel real-time DBS algorithm based on CZT has been proposed. The proposed algorithm has two advantages: (1) it simplify the modified DBS processing by circumventing Doppler centroid estimation, which makes it applicable to real-time project; (2) resolution loss caused by Doppler centroid estimation error is avoided.

At last, based on real SAR data of WAS GMTI mode, a GMTI and parameter estimation algorithm has been presented based on multiple revisits. By using the multiple revisits of the same observation area, the location of the moving target is different in each image, and the motion parameters can be estimated from the location changes. However, this algorithm still faces the shortcomings that it is not suited for real-time processing.

References

1. Wilden H, Brenner AR (2010) The SAR/GMTI airborne radar PAMIR: technology and performance. In: IMS 2010, pp 534–537
2. Wei S (2005) Moving target indication and imaging in airborne SAR. Huazhong University of Science and Technology
3. Song D (2007) DBS technique in mechanic scan radar. Xidian University
4. Mao S, Li S, Huang Y et al (2000) Study of real-time image by DBS on airborne PD radar. ACTA Electronica Sin 28(3):32–34
5. Zhang Z (1991) Several issues in DBS theory and applications. Mod Radar 13(2):1–12
6. Wei H, Li M (2009) Study on keeping sharpening ratio constant technique in DBS mode of mechanical scanning radar. Fire Control Radar Technol 28(2):14–18
7. Qiu T, Du Z, Zhang T (2005) Technique of Doppler beam sharpening for enhancing radar azimuth resolution. Fire Control Radar Technol 34(2):17–20
8. Liu Y, Wu S (2005) Combination algorithm for DBS sub-images in scanning mode. Radar Sci Technol 2:91–95
9. Li Y (2006) DBS imaging and moving target indication in mechanical scanning radar. Xidian University
10. Ding Y, Zhang B (2008) Arithmetic estimation of DBS Doppler center. J Geomatics Sci Technol 25(3):176–182
11. Wang H, Wei S, Sun W (2005) High resolution DBS imaging and moving target trajectory processing. Chin J Radio Sci 20(5):637–641
12. Zhao Y, Wang Y (2008) A new GMTI method based on Chirp-Z Transform for distributed satellites interferometric SAR. J Electron Inf Technol 30(5):1056–1059
13. Rabiner LR, Schafer RW, Rader CM (1969) The chirp z-transform algorithm. IEEE Trans Audio Electroacoust AU 17(2):86–92

Chapter 5
GMTI and GMTIm in FMCW SAR

Abstract This chapter mainly discusses the GMTI and GMTIm algorithms in FMCW SAR. The nonlinearity correction and GMTI algorithms in FMCW SAR are studied. Then, the differences between a FMCW SAR and a pulse SAR are analyzed, and additional phase error of FMCW SAR is calculated. Furthermore, a nonlinearity correction algorithm based on Homomorphic Deconvolution is proposed, and an along-track moving target indication algorithm is presented. By using the additional phase error, the GMTI performance of FMCW SAR is improved.

5.1 Introduction

Miniaturization is a developing tendency of the SAR system. A novel FMCW SAR system with small size, light weight, low power-cost, and high range resolution is developed by applying FMCW technology into SAR system [1]. Since the concept of FMCW SAR was proposed by London University in 1988, U.K. [2], Japan, Netherlands, U.S., German and China have all devoted to this domain. The FMCW SAR system from TUDelft University [3], MiSAR from German, MicroSAR and NanoSAR from the U.S. are all practical FMCW SAR systems. The advantages of FMCW SAR make it highly suitable for small platforms, such as a UAV, and make it useful in short-range military surveillance and target indication. The low-cost character also promises it a broad application prospect in civil area. Therefore, FMCW SAR is a hotspot in SAR system research.

In FMCW SAR systems, the radar transmits chirp signals during the entire PRI, and the deramping operation is used in the received signal to reduce the burden of A/D [4]. However, the unique transmit and receive operations make the signal model of FMCW SAR different from that of a pulse SAR. The stop-and-go approximation is not valid in FMCW SAR, and the RVP is induced by the deramping operation [5].

In practical applications, the continuous transmit and receive also induced the crosstalk of the transmitted and received signals. This problem is also solved by separated transmitter and receiver, thus is not discussed in this chapter [6]. The

J. Yang, *Study on Ground Moving Target Indication and Imaging Technique of Airborne SAR*, Springer Theses, DOI 10.1007/978-981-10-3075-8_5

large time-bandwidth product signal of FMCW SAR cannot maintain strictly linear, the existence of nonlinearity in transmitted signal severely limits the performance of FMCW SAR imaging. After deramping operation, the transmitted nonlinearity is overlapped with the received nonlinearity in beat signal, making the nonlinearity problem more complicated, especially for large distance [7]. Therefore, nonlinearity must be corrected either in signal production stage or during imaging process.

The small size and light weight characters make FMCW SAR high suitable for military applications, such as missile-borne seeker. Therefore, the research on the GMTI and GMTIm algorithms in FMCW SAR are important. The principle of FMCW SAR makes its signal echo different from that of a pulse SAR. This chapter establishes the echo model of a moving target in FMCW SAR, and analyzes the echo phase errors in detail. These phase errors are compensated in traditional FMCW SAR GMTI and GMTIm algorithms as interference, while a novel algorithm is proposed to indicate along-track moving targets based on the additional phase error.

This chapter mainly focuses on the nonlinearity correction, GMTI and GMTIm of FMCW SAR. Section 5.2 analyzes the echo model of FMCW SAR. In Sect. 5.3, the nonlinearity problem is analyzed, and a correction algorithm based on homomorphic deconvolution is proposed. An along-track moving target indication algorithm is presented in Sect. 5.4. In Sect. 5.5, we address the conclusion.

5.2 Principle of FMCW SAR and Signal Model

5.2.1 Principle of FMCW SAR

The range resolution of SAR is determined by the bandwidth of the transmitted signal. FMCW is an applicable technique to improve the range resolution of SAR without a high frequency modulation rate. In FMCW SAR systems, the radar transmits chirp signals during the entire PRI. In pulse SAR systems, the pulse duration is on the order of a few microseconds, whereas the pulse duration is on the order of milliseconds in FMCW SAR systems. Therefore, with the long pulse duration, the bandwidth of the transmitted signal in FMCW SAR systems is increased, and the range resolution is improved, as shown in Fig. 5.1.

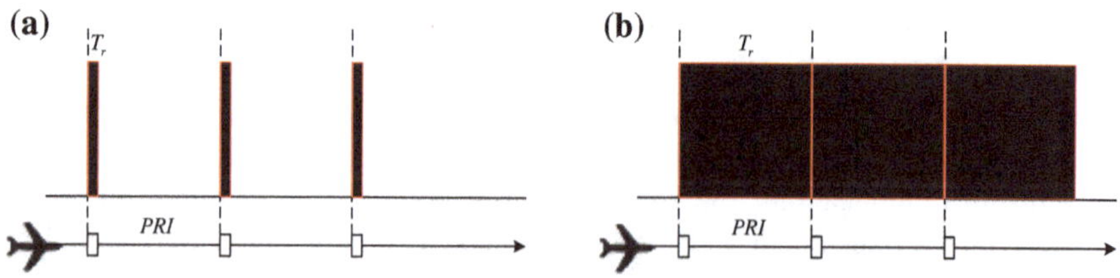

Fig. 5.1 Comparison of signals in pulse SAR and FMCW SAR. **a** Transmitted signal in pulse SAR. **b** Transmitted signal in FMCW SAR

In pulse SAR systems, the stop-and-go approximation is commonly used, and the instantaneous slant range from the antenna to the target is assumed constant during the pulse duration. In the case of FMCW SAR, however, this approximation is inaccurate due to the long pulse duration.

Moreover, given that the transmitted bandwidth is large in FMCW SAR systems, the deramping operation is used to reduce the range sampling rate. The deramping operation is shown in Fig. 5.2. In Fig. 5.2, the transmitted signal is represented by the black line, and the received signals from near range and far range are represented by red and blue lines, respectively. The reference signal is represented by the green line. The RVP is induced by the deramping operation, which needs to be corrected during the imaging process [8].

It can also be noted from Fig. 5.2 that after the deramping operation, the deramped signals are monochromatic signals. Set the near range signal as an example, the deramped frequency f_{near} can be expressed as

$$f_{near} = \frac{B_r}{PRI} \frac{2(R_{near} - R_{ref})}{c} \tag{5.2.1}$$

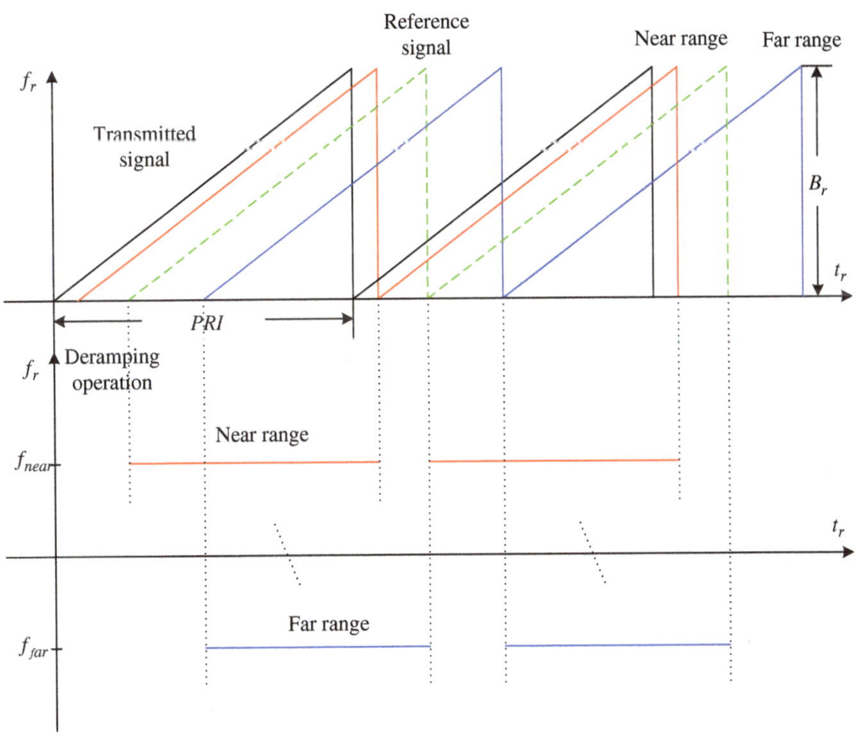

Fig. 5.2 Principle of the deramping operation in FMCW SAR

where R_{near} is the near range, and B_r is the bandwidth of the transmitted signal. The second-order phase error of the beat signal is compensated, and the range compression of FMCW SAR can be accomplished by range FFT [9].

5.2.2 Signal Model and Analysis

The geometry of stationary targets in FMCW SAR is the same as a pulse SAR, so the geometry is the same as Fig. 2.1. However, as analyzed in Sect. 5.2.1, the stop-and-go approximation is not valid in FMCW SAR, so the instantaneous slant range between stationary target P and the platform at azimuth time t_a can be expressed as

$$R(t_a, t_r) = \sqrt{R_0^2 + V_a^2(t_a + t_r)^2} \tag{5.2.2}$$

It can be noted that the instantaneous slant range is relative to the range time. After Taylor series expansion, Eq. (5.2.2) can be approximated as

$$R(t_a, t_r) \approx R_0 + \frac{V_a^2}{2R_0}t_a^2 + \frac{V_a^2}{R_A}t_a t_r \tag{5.2.3}$$

where $R_A = \sqrt{R_0^2 + V_a^2 t_a^2}$, and is approximated as $R_A \approx R_0$. Compare Eq. 5.2.3 with Eq. 2.2.2, a range-azimuth coupling term is added in FMCW SAR. Since the azimuth information is hidden in the slant range, the echo phase of FMCW SAR is different from that of pulse SAR.

Neglect the impacts of signal amplitude and window function, the transmitted signal of FMCW SAR can be expressed as

$$s_t(t_r) = \exp\left(j2\pi\left(f_0 t_r + \frac{1}{2}K_r t_r^2\right)\right) \tag{5.2.4}$$

And the received signal can be expressed as a copy of the transmitted signal with a time-delay:

$$s_r(t_a, t_r) = \exp\left(j2\pi\left(f_0\left(t_r - \frac{2R(t_a, t_r)}{c}\right) + \frac{1}{2}K_r\left(t_r - \frac{2R(t_a, t_r)}{c}\right)^2\right)\right) \tag{5.2.5}$$

Before the received signal is sampled by A/D, the received signal must be deramped by the reference signal. The reference signal is the echo from a known range R_{ref}, which is represented as

$$s_{ref}(t_r) = \exp\left(j2\pi\left(f_0\left(t_r - \frac{2R_{ref}}{c}\right) + \frac{1}{2}K_r\left(t_r - \frac{2R_{ref}}{c}\right)^2\right)\right) \tag{5.2.6}$$

Send Eqs. 5.2.5 and 5.2.6 into a conjugate multiplier, set $R_{ref} = 0$, and the beat signal can be presented as

$$s_{if}(t_a, t_r) = \exp\left(j\frac{4\pi}{\lambda}R(t_a, t_r)\right)\exp\left(j\frac{4\pi K_r}{c}R(t_a, t_r)t_r\right)\exp\left(-j\frac{4\pi K_r}{c}R(t_a, t_r)^2\right) \tag{5.2.7}$$

The last term in Eq. 5.2.7 is the RVP term, and must be corrected in the imaging process. After the RVP correction, substitute Eq. 5.2.3 into Eq. 5.2.7 and neglect the higher terms of t_r, it yields

$$\begin{aligned} s_{if}(t_a, t_r) = {} & \exp\left(j\frac{4\pi}{\lambda}R_0\right)\exp\left(j\frac{4\pi K_r}{c}R_0 t_r\right)\exp\left(j\frac{2\pi V_a^2}{\lambda R_0}t_a^2\right) \\ & \exp\left(j\frac{2\pi K_r V_a^2}{cR_0}t_a^2 t_r\right)\exp\left(j\frac{4\pi}{\lambda}\frac{V_a^2}{R_A}t_a t_r\right) \end{aligned} \tag{5.2.8}$$

The first term of Eq. 5.2.8 is a constant term. The second term is the first-order term of range time, which indicates that the range compression can be accomplished by FFT. The third and fourth terms are the azimuth phase and the range curvature. These four terms are all caused by the movement of the radar platform, which are the same as pulse SAR.

The last term of Eq. 5.2.8 is the RWM induced by the movement of the radar platform within a PRI. This term is caused by the invalidation of the stop-and-go approximation, and is ignored in the pulse SAR. Therefore, the imaging algorithms in pulse SAR is not valid in FMCW SAR anymore, and the additional range walk must be corrected.

5.3 Nonlinearity Correction of FMCW SAR

The continuous transmit and receive signals during the entire PRI of FMCW SAR induces the problem that the phase linearity cannot be maintained. The nonlinearity of the transmitted signal will severely affect the performance of the range compression, and furthermore deteriorate the image resolution.

The nonlinearity correction is the prior issue in FMCW SAR signal processing, since the range compression is the foundation of the following signal processing. Both hardware and software solutions have been proposed to solve this problem. Hardware approaches are proposed to eliminate nonlinearity in signal generating stage, such as using pre-distortion techniques to compensate for nonlinearity in VCOs response characteristics, using a tuning voltage converter to correct varactor tuning curves, direct digital synthesizer [10–12], and other compensation circuits.

However, these techniques are either expensive or increasing the complexity of the system. Moreover, the performances of hardware approaches suffer from external conditions. Software approaches can be implemented without restrictions of hardware methods. Jiang et al. [13] suggested an adaptive sampling method to correct nonlinearity in FMCW radar. Kulpa et al. [14] presented a nonlinearity estimation algorithm with curve fitting. Nevertheless, both algorithms are based on specific approximated nonlinearity models, which limit their generalities in practical FMCW SAR applications. Meta et al. [15] proposed a transmitted nonlinearity estimation algorithm regarding the nonlinearity in beat signal as the derivative of transmitted nonlinearity. By performing integral operation on beat signal with known propagation delay, the transmitted nonlinearity can be estimated. However, this algorithm is inherently biased since the derivative approximation is used, and its performance degrades rapidly as time delay increases.

In this section, the nonlinearity problem is analyzed by signal modeling, and the existing method is introduced, and its limitations are discussed. Moreover, the proposed algorithm based on homomorphic deconvolution and the correction strategy are presented, and simulation results are shown to demonstrate the validity of the proposed algorithm.

5.3.1 Analysis of the Nonlinearity Problem

The ideal transmitted signal, received signal and beat signal without nonlinear phase error can be expressed respectively as

$$s_{t\_lin}(t_r) = \exp\left(j2\pi\left(f_0t_r + \frac{1}{2}K_rt_r^2\right)\right) \tag{5.3.1}$$

$$s_{r\_lin}(t_r, \tau) = \exp\left(j2\pi\left(f_0(t_r - \tau) + \frac{1}{2}K_r(t_r - \tau)^2\right)\right) \tag{5.3.2}$$

$$s_{if\_lin}(t_r, \tau) = \exp\left(j2\pi\left(f_0\tau + K_rt_r\tau - \frac{1}{2}K_r\tau^2\right)\right) \tag{5.3.3}$$

where τ is the time delay. As shown in Eq. 5.3.3, the signal after the deramping operation is a signal-frequency signal of t_r, and the last term is the RVP term.

If the nonlinear phase error exists in the system, the transmitted signal is added with a nonlinear phase error. Suppose the nonlinearity error is $\varepsilon(t)$, the transmitted signal with nonlinear phase error can be expressed as

$$s_t(t_r) = \exp\left(j2\pi\left(f_0t_r + \frac{1}{2}K_rt_r^2 + \varepsilon(t_r)\right)\right) = s_{t\_lin}(t_r)s_\varepsilon(t_r) \tag{5.3.4}$$

The received signal can be expressed as

$$\begin{aligned} s_r(t_r, \tau) &= \exp\left(j2\pi\left(f_0(t_r - \tau) + \frac{1}{2}K_r(t_r - \tau)^2 + \varepsilon(t_r - \tau)\right)\right) \\ &= s_{r\_lin}(t_r, \tau)s_\varepsilon(t_r - \tau) \end{aligned} \tag{5.3.5}$$

And the beat signal can be expressed as

$$\begin{aligned} s_{if}(t_r, \tau) &= \exp\left(j2\pi\left(f_0\tau + K_r t_r \tau + \frac{1}{2}K_r\tau^2 + \varepsilon(t_r) - \varepsilon(t_r - \tau)\right)\right) \\ &= s_{if\_lin}(t_r, \tau)s_\varepsilon(t_r)s_\varepsilon^*(t_r - \tau) \end{aligned} \tag{5.3.6}$$

It can be noted that the transmitted and received nonlinearities are overlapped in beat signal, and the nonlinearity term in Eq. 5.3.6 differs as time delay changes. If the time delay is small, most of the nonlinearities from the transmitted and received signals are eliminated, while the nonlinear phase error increases with the time delay in the beat signal.

$$s_\varepsilon(t_r)s_\varepsilon^*(t_r - \tau) = \exp(j2\pi(\varepsilon(t_r) - \varepsilon(t_r - \tau))) \tag{5.3.7}$$

The phase in Eq. 5.3.7 can be considered as the difference of the transmitted nonlinearity and its attenuated copy. Hence, to correct nonlinearity of beat signal, the transmitted nonlinearity must be estimated first. The estimation precision of transmitted nonlinearity determines the performance of nonlinearity correction of FMCW SAR.

5.3.2 *Principle of Derivative Algorithm*

To obtain $s_\varepsilon(t)s_\varepsilon(t - \tau)^*$ from the beat signal, a controllable delay line is usually involved in nonlinearity correction algorithms. The delay line technique provides a controllable time delay τ_{ref} of the original signal without changing its amplitude. Therefore, the nonlinear error phase can be calculated from Eq. 5.3.8 as

$$\Delta\Phi(t_r, \tau_{ref}) = \varepsilon(t_r) - \varepsilon(t_r - \tau_{ref}) \tag{5.3.8}$$

Meta et al. [15] proposed an estimation algorithm which is genetic to all nonlinearity models. This algorithm regards the difference of $\varepsilon(t_r)$ in beat signal as the derivative of $\varepsilon(t_r)$ in the case that the time delay τ_{ref} is far smaller than *PRI*. Therefore, the nonlinear phase error can be approximated as

$$\Delta\Phi(t_r, \tau_{ref}) = \varepsilon(t_r) - \varepsilon(t_r - \tau_{ref}) \approx \tau_{ref}\varepsilon'(t_r) \tag{5.3.9}$$

Thus, $\varepsilon(t_r)$ can be estimated as

$$\varepsilon(t_r) \approx \int_0^{PRI} \left(\Delta\Phi\left(t_r, \tau_{ref}\right)/\tau_{ref}\right) dt_r \qquad (5.3.10)$$

The algorithm in [15], referred to as derivative algorithm, is an efficient method with simple processing steps. If τ_{ref} is far smaller than *PRI*, it will give an accurate estimation of transmitted nonlinearity. However, it is inherently a biased algorithm since it is based on the derivative approximation, which makes it inapplicable in high resolution imaging. Moreover, estimation performance decreases sharply as time delay increases.

5.3.3 Nonlinearity Correction Algorithm Based on Homomorphic Deconvolution

Bogert et al. [16] proposed the concept of homomorphic deconvolution. Its basic idea is: as to signals with echo interruption, the logarithm of its power spectrum contains an additive periodic component, which appears a peak at the time delay of echo signal after IFFT. The original signal can be reverted by eliminating this peak. The time domain after processing IFFT on the logarithm is defined as Cepstrum domain. Since time domain and Cepstrum domain are homomorphic about convolution and addition, the algorithm to revert a time domain signal by filtering in Cepstrum domain is defined as homomorphic deconvolution. The concept of homomorphism is: set A and set B are defined as homomorphic about transforms Δ and *O*, if and only if there exists a surjection from A to B, that for any element $a_1, a_2 \in A$ and $b_1, b_2 \in B$, if $a_1 \rightarrow b_1$ and $a_2 \rightarrow b_2$, then $a_1 \Delta a_2 \rightarrow b_1 O b_2$ [17].

Similar as multipath interference in communication signal processing, homomorphic deconvolution can be used to revert the transmitted nonlinearity in FMCW SAR. Consider $\varepsilon(t_r)$ as the original signal, and $\Delta\Phi\left(t_r, \tau_{ref}\right)$ as the interferential signal, then $\Delta\Phi\left(t_r, \tau_{ref}\right)$ can be regarded as a convolution of $\varepsilon(t_r)$ and a δ function series represented by $p\left(t_r, \tau_{ref}\right)$

$$\Delta\Phi\left(t_r, \tau_{ref}\right) = \varepsilon(t_r) \otimes p\left(t_r, \tau_{ref}\right) \qquad (5.3.11)$$

where $p\left(t_r, \tau_{ref}\right) = \delta(t_r) - \delta\left(t_r - \tau_{ref}\right)$. Perform FFT on Eq. 5.3.11, the spectrum of $\Delta\Phi\left(t_r, \tau_{ref}\right)$ can be obtained as

$$\Delta\Phi(\omega) = \varepsilon(\omega) \cdot \left[1 - \exp\left(-j\pi\tau_{ref}\right)\right] \qquad (5.3.12)$$

Perform logarithm transform on Eq. 5.3.12, then Eq. 5.3.12 can be rewritten as

$$\ln \Delta\Phi(\omega) = \ln \varepsilon(\omega) + \ln\left[1 - \exp\left(-j\pi\tau_{ref}\right)\right] \tag{5.3.13}$$

Expand term $\ln\left[1 - \exp\left(-j\pi\tau_{ref}\right)\right]$ in Eq. 5.3.13 into a Taylor series

$$\ln\left[1 - \exp\left(-j\pi\tau_{ref}\right)\right] = -\sum_{k=1}^{\infty} \frac{\exp\left(-j\pi k\tau_{ref}\right)}{k} \tag{5.3.14}$$

Substitute Eq. 5.3.14 into Eq. 5.3.13, and perform IFFT on Eq. 5.3.14, it yields

$$\Delta\hat{\Phi}\left(t_r, \tau_{ref}\right) = \hat{\varepsilon}(t_r) - \sum_{k=1}^{\infty} \frac{1}{2\pi k} \delta\left(t_r - k\tau_{ref}\right) \tag{5.3.15}$$

Equation 5.3.15 is the expression of nonlinear error phase in beat signal in Cepstrum domain. δ series should be eliminated to get the transmitted nonlinearity $\hat{\varepsilon}(t_r)$. Since time delay τ_{ref} is known, one direct method is to construct a δ series according to the second term on the right of (5.3.15). Thus, δ series can be eliminated by subtracting the constructed δ series function. This method is a direct and ideal method that has an optimal performance in estimating the transmitted nonlinear error because the energy of $\hat{\varepsilon}(t_r)$ is not changed after the subtraction.

However, the δ series subtraction method can only be used in simulations. δ function is a pulse signal with infinite amplitude at one frequency spot in analog system. But in digital system and computer simulation, it is represented by a signal with amplitude of 1 at one frequency spot. Therefore, there may be differences between the simulated δ series and δ pulses in real data, and it will induce estimation error when using δ series subtraction method in practical application.

In order to accomplish the same goal of δ series subtraction method, a comb notch filter is generated in Cepstrum domain to remove δ series. Comb notch filter is a filter with amplitude of zero at the points of $t = k\tau_{ref}$ and one at other points. The width of the filter is the same as the signal. Given that the time-width of the signal is T_r and the length of the signal is N_r, the width corresponding to time delay τ_{ref} can be calculated as

$$N_\tau = round\left(\frac{\tau_{ref}}{T_r} N_r\right) \tag{5.3.16}$$

Hence, the number of notches is $N_r/N_\tau - 1$. The waveform of a comb notch filter is shown in Fig. 5.3. After Cepstrum filtering, $\hat{\varepsilon}(t_r)$ can be reverted from the interferential signal.

In the case that the energy of the phase error concentrates at $t = k\tau_{ref}$, energy of the phase error will be filtered out by the comb notch filter as well as δ series at $t = k\tau_{ref}$. However, since the width of the notch is only 1–2 spots, the majority of phase error energy can still be reserved after the filtering. Therefore, comb notch filter method is also effective in the case that energy of the original phase error

Fig. 5.3 Waveform of the comb notch filter

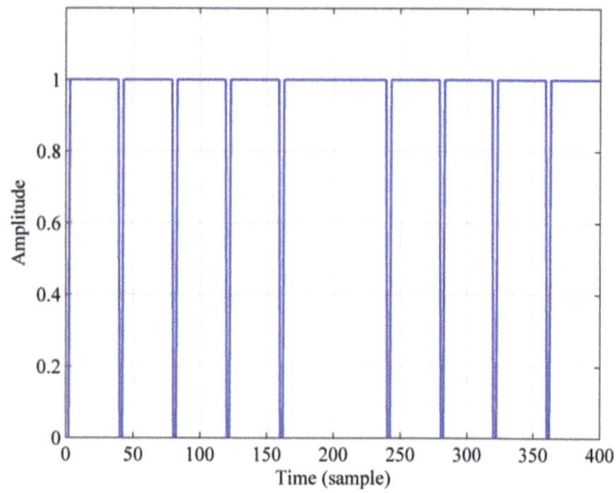

concentrates at $t = k\tau_{ref}$. Moreover, comb notch filter is easy to be built comparing with δ series subtraction method, and it is reliable in practical use since filtering technique has been widely used in signal processing field. Therefore, comb notch filter is used in our algorithm to eliminate δ series in Cepstrum domain.

Throughout the whole algorithm, no approximations are used, and no specific model of $\varepsilon(t_r)$ is required. Therefore, this algorithm can give a theoretically unbiased estimation of $\varepsilon(t_r)$. Moreover, compared with derivative algorithm, the proposed algorithm has no limitation of time delay τ_{ref}, thus the effectiveness of the proposed algorithm in practical application is further strengthened. The flowchart of the proposed algorithm is illustrated in Fig. 5.4.

The nonlinearities in the beat signal vary among each range cell. This will make the correction process complicated and not applicable to real-time tasks. To address this problem, a strategy is proposed to eliminate the dependence of nonlinearity on τ using RVP removal. After RVP removal, the beat signal is independent of τ, and the received nonlinearity can be corrected by a unique correction function

$$s_{\varepsilon\_RVP} = F^{-1}\left\{F\{s_\varepsilon(t_r)\}\exp\left(j\pi\frac{f_r^2}{K_r}\right)\right\} \approx \exp\{j\pi[\varepsilon_{RVP}(t_r)]\} \tag{5.3.17}$$

where $F\{\cdot\}$ and $F^{-1}\{\cdot\}$ denote FFT and IFFT, respectively.

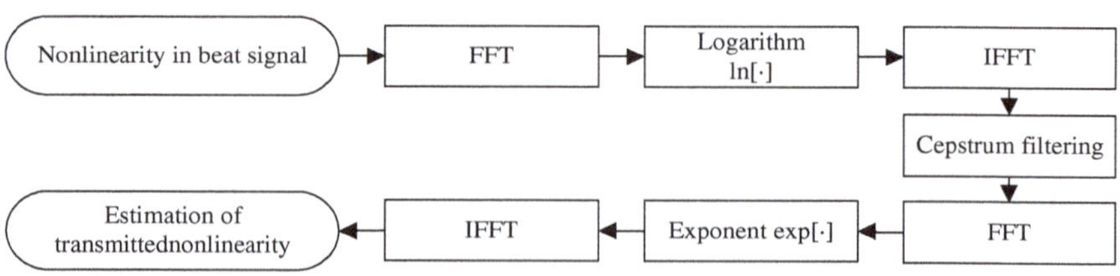

Fig. 5.4 Flowchart of the proposed estimation algorithm

Figure 5.5 shows the signal of each steps of the strategy. Figure 5.5a shows the beat signal with nonlinear error in time-frequency domain. The larger the time delay is, the larger the nonlinearity will be. Figure 5.5b shows that after transmitted nonlinearity correction, there remains received nonlinearity that dependent on time delay. Figure 5.5c shows that using RVP removal, the delay-dependence of beat signal is eliminated. After received nonlinearity correction, the ideal time-frequency characters of beat signal are shown in Fig. 5.5d. Beat signal is reverted to a monochromatic signal after the strategy.

Figure 5.6 illustrates the flowchart of FMCW SAR nonlinearity correction strategy combining with homomorphic deconvolution.

To give a full-scale simulation of the proposed algorithm, quadratic, cubic, and sinusoid nonlinearity errors are simulated in this section. In addition, algorithm proposed in [15] is also operated as a comparison.

Figure 5.7a, c, e show the performances of the proposed and derivative algorithms with quadratic, cubic, and sinusoid nonlinear errors, respectively. For each nonlinearity model, the proposed algorithm provides more accurate nonlinearity estimations than the derivative algorithm. Moreover, derivative algorithm introduces erroneous estimations at the beginning of the signal, which should be

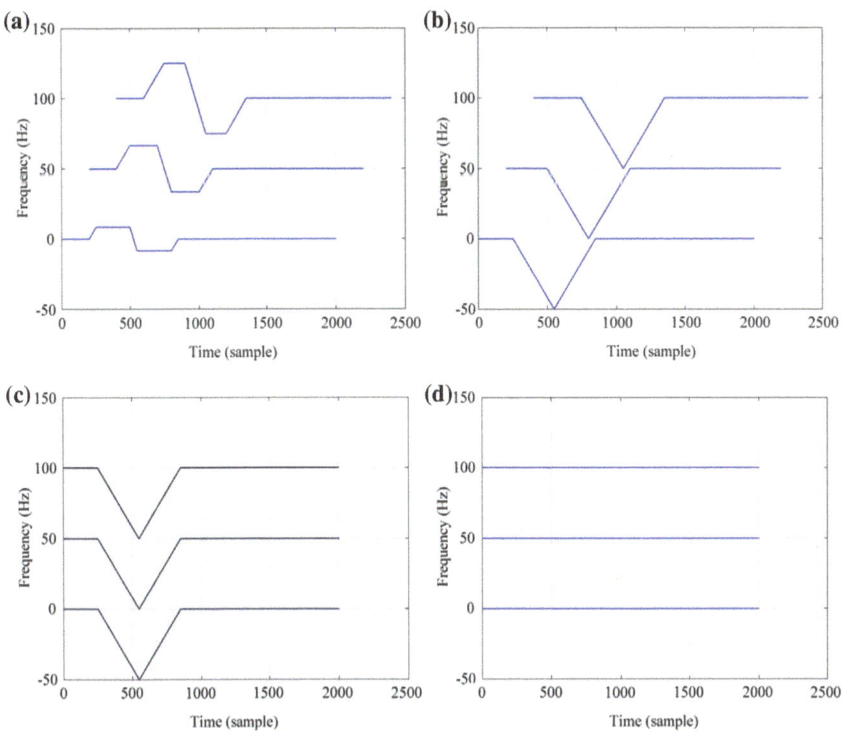

Fig. 5.5 FMCW SAR nonlinearity correction strategy. **a** Beat signal with nonlinear error. **b** After transmitted nonlinearity correction. **c** After RVP removal. **d** After received nonlinearity correction

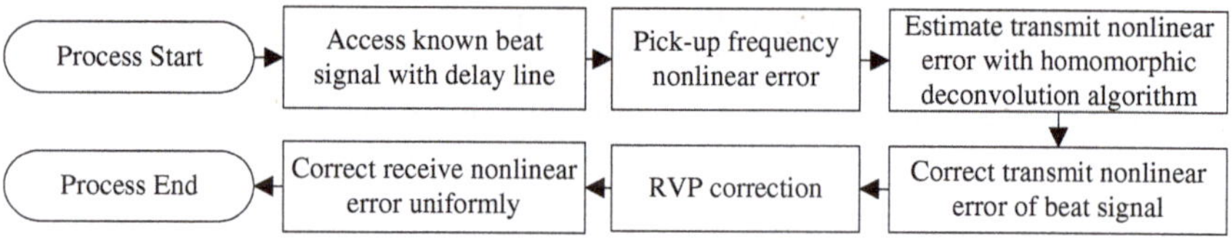

Fig. 5.6 Flowchart of FMCW SAR nonlinearity correction with proposed algorithm

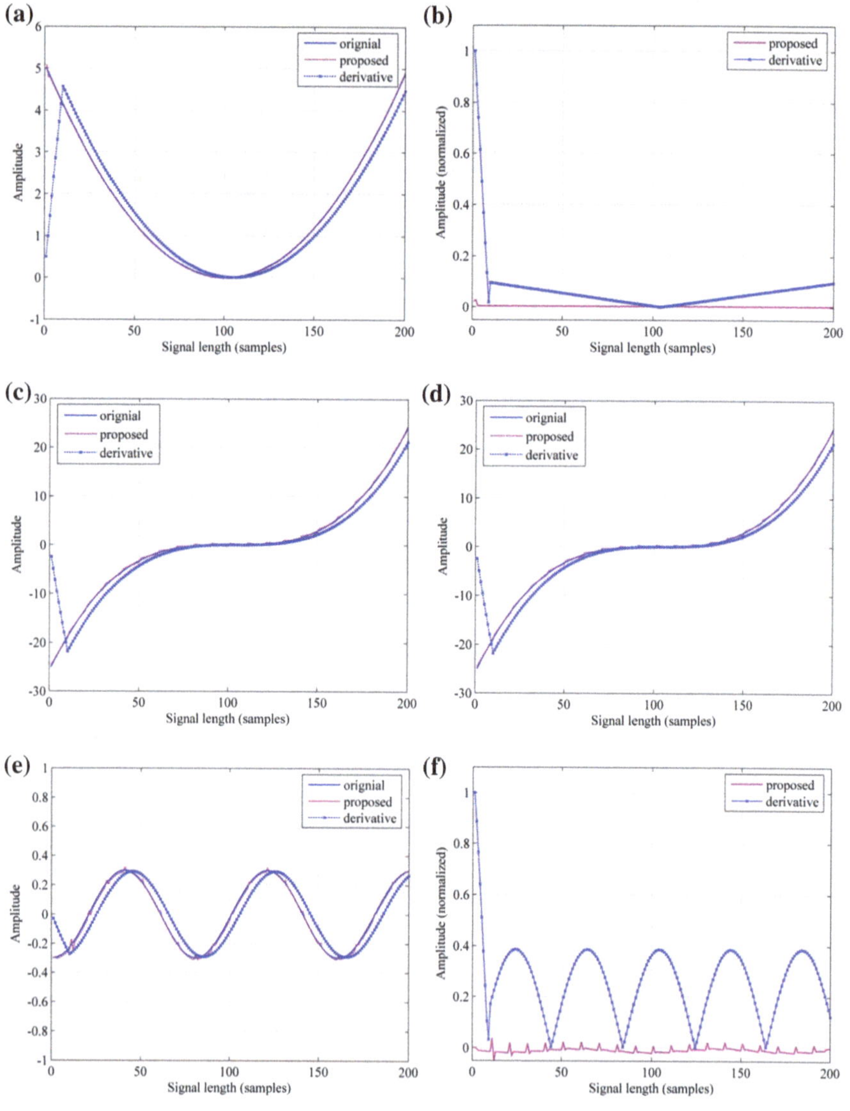

Fig. 5.7 Comparison of the estimation accuracy of both algorithms. **a** Quadratic nonlinearity simulation. **b** Estimation error of quadratic errors. **c** Cubic nonlinearity simulation. **d** Estimation error of cubic errors. **e** Sinusoid nonlinearity simulation. **f** Estimation error of sinusoid errors

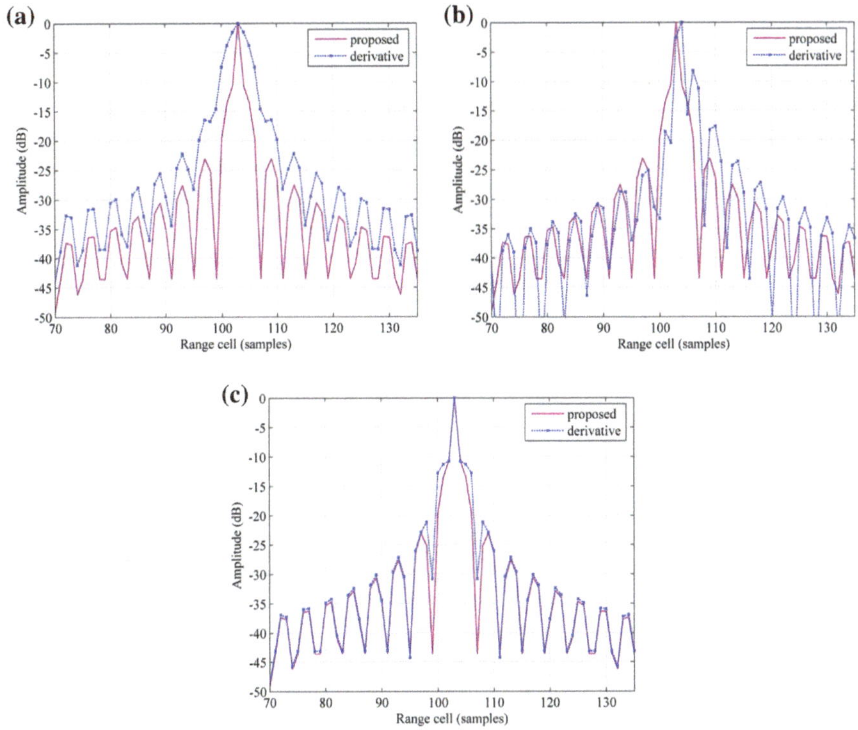

Fig. 5.8 Comparison of the imaging qualities of both algorithms. **a** Quadratic nonlinearity simulation. **b** Cubic nonlinearity simulation. **c** Sinusoid nonlinearity simulation

abandoned. To make a more clear comparison of both algorithms, the estimation errors of each model are shown in Fig. 5.7b, d, f. In addition, in Fig. 5.7e, f, a sinusoidal phase error with oscillation frequency τ_{ref} is simulated. The energy of the phase error is concentrated at frequency spots $t = k\tau_{ref}$ in this simulation. It can be concluded from the results that the proposed algorithm still has a convincing performance in this case.

Figure 5.8a shows the result of range compression of FMCW SAR with quadratic nonlinearity. Quadratic nonlinearity causes main lobe broadening, and the imaging quality of derivative algorithm is worse than that of the proposed algorithm. Figure 5.8b shows the imaging result comparison with cubic nonlinearity. Since nonlinearity is not completely removed by derivative algorithm, its side-lobes are raised and appear to be unsymmetrical. Figure 5.8c illustrates the comparison results with sinusoid nonlinearity, in which the proposed algorithm also provides a better resolution. Since azimuth resolution is not affected by nonlinearity problem, simulation in Fig. 5.8 is adequate to prove the imaging resolution improvement of the proposed algorithm.

Another simulation is used to verify the robustness of the proposed algorithm. Quadratic nonlinearity model is used to represent all models since the generality

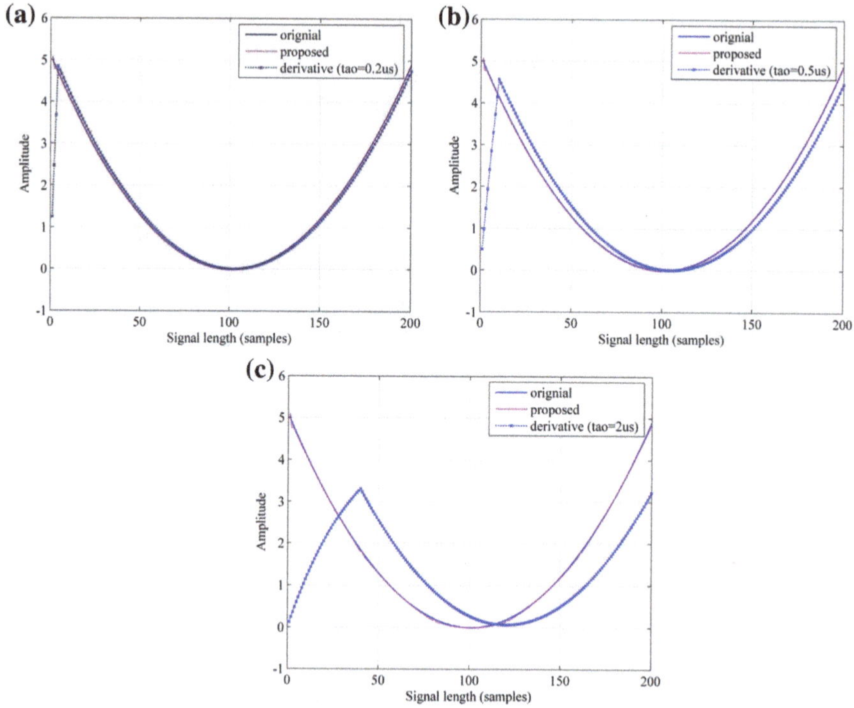

Fig. 5.9 Comparison of estimation accuracies with different delay. **a** Comparison results with delay of 0.2 μs. **b** Comparison results with delay of 0.5 μs. **c** Comparison results with delay of 2 μs

character has been demonstrated. Set *PRI* as 10 μs, and τ_{ref} gets the values of 0.2, 0.5 and 2 μs. Figure 5.9a shows that when $\tau_{ref} = 0.2$ μs, the proposed and derivative algorithm both provide accurate estimations of the nonlinearity error. When $\tau_{ref} = 0.5$ μs, the derivative algorithm induces an considerable estimation error in Fig. 5.9b. In Fig. 5.9c, since $\tau_{ref} = 2$ μs, the estimation of derivative algorithm obviously departs from the actual value. However, the proposed algorithm performs a robust estimation with high accuracy in all three simulations. Therefore, the robustness of the proposed algorithm is proved.

5.4 Along-Track Moving Target Indication and Imaging in FMCW SAR

5.4.1 Signal Model of Moving Targets in FMCW SAR

The geometry of a moving target in FMCW SAR is the same as that in a pulse SAR system, as shown in Figs. 2.2 and 2.3. To briefly compare the additional phase error of FMCW SAR, the accelerations are neglected, and the instantaneous slant range of a moving target in FMCW SAR can be expressed as

$$R(t_a, t_r) = \sqrt{[R_0 - V_r(t_a + t_r)]^2 + (V_a - V_y)^2 (t_a + t_r)^2} \tag{5.4.1}$$

Perform Taylor expansion of Eq. 5.4.1, it can be approximated as

$$R(t_a, t_r) \approx R_0 - V_r t_a + \frac{(V_a - V_y)^2}{2R_0} t_a^2 + \frac{V_r^2 + (V_a - V_y)^2}{R_A} t_a t_r - \frac{R_0}{R_A} V_r t_r \tag{5.4.2}$$

where $R_A = \sqrt{(R_0 - V_r t_a)^2 + (V_a - V_y)^2 t_a^2}$. Compare Eq. 5.4.2 with Eq. 2.3.4, two additional phase terms exists in Eq. 5.4.2. Substitute Eq. 5.4.2 into Eq. 5.2.7 and perform RVP correction, it yields

$$\begin{aligned} s_{if}(t_a, t_r) = {} & \exp\left(j\frac{4\pi}{\lambda} R_0\right) \exp\left(j\frac{4\pi K_r}{c} R_0 t_r\right) \exp\left(-j\frac{4\pi}{\lambda} V_r t_a\right) \\ & \cdot \exp\left(j\frac{2\pi (V_a - V_y)^2}{\lambda R_0} t_a^2\right) \exp\left(-j\frac{4\pi}{\lambda}\frac{R_0}{R_A} V_r t_r\right) \exp\left(j\frac{4\pi}{\lambda}\frac{V_r^2 + (V_a - V_y)^2}{R_A} t_a t_r\right) \\ & \cdot \exp\left(-j\frac{4\pi K_r}{c} V_r t_a t_r\right) \exp\left(j\frac{2\pi K_r (V_a - V_y)^2}{c R_0} t_a^2 t_r\right) \end{aligned} \tag{5.4.3}$$

The fifth term in Eq. 5.4.3 is the range location shift induced by cross-track velocity. A moving target is dis-located in the range direction in a stationary FMCW SAR image. The sixth term in Eq. 5.4.3 is the FMCW system. The seventh term in Eq. 5.4.3 is also a range walk, which is induced by the cross-track velocity.

Many researchers have developed algorithms to indicate and focus moving targets in FMCW SAR by using the GMTI and GMTIm algorithms and correct the additional phase errors [18, 19]. However, we discovered that the moving targets can be indicated from the additional phase errors, and the motion parameters can be extracted from them.

5.4.2 *Along-Track Moving Target Indication Algorithm in FMCW SAR*

The geometry of an along-track moving target in a FMCW SAR system is illustrated in Fig. 5.10. In Fig. 5.10, the instantaneous slant range can be expressed as

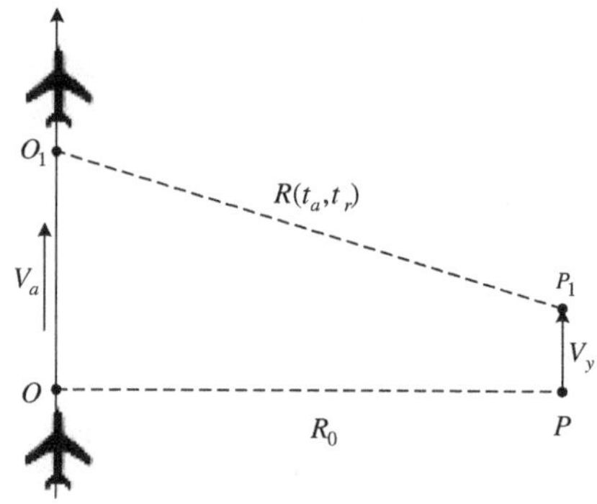

Fig. 5.10 Geometry of an along-track moving target in FMCW SAR

$$R(t_a,t_r)=\sqrt{R_0^2+(V_a-V_y)^2(t_a+t_r)^2}\approx R_0+\frac{(V_a-V_y)^2}{2R_0}t_a^2+\frac{(V_a-V_y)^2}{R_0}t_at_r \tag{5.4.4}$$

Substitute Eq. 5.4.4 into Eq. 5.2.7, the echo of the along-track moving target can be presented as

$$\begin{aligned} s_{if}(t_a,t_r)=&\exp\left(j\frac{4\pi}{\lambda}R_0\right)\exp\left(j\frac{4\pi K_r}{c}R_0t_r\right)\exp\left(j\frac{2\pi(V_a-V_y)^2}{\lambda R_0}t_a^2\right)\\ &\cdot\exp\left(j\frac{4\pi}{\lambda}\frac{(V_a-V_y)^2}{R_0}t_at_r\right)\exp\left(j\frac{2\pi K_r(V_a-V_y)^2}{cR_0}t_a^2t_r\right) \end{aligned} \tag{5.4.5}$$

The fourth term of Eq. 5.4.5 is the range walk induced by FMCW system, which is determined by the along-track velocity. The additional range walk is often removed as a phase error in the existing FMCW SAR GMTIm algorithms. However, it is actually a favorable factor for the moving target indication. After range compression and RCMC, the trajectories of the stationary targets are straight lines that are parallel to the azimuth direction, whereas the trajectories of the along-track moving targets are not parallel to the azimuth direction as a result of the residual range curvature and the additional range walk.

Hough transform was proposed by P.V.C. Hough (1962) for the image characteristic detection. It is widely used in digital image processing, especially in dealing with the straight line indication and parameter estimation. After the Hough transform, the RCM trajectories in the scene are transformed into peaks. In the Hough transform domain, the amplitudes and locations of the peaks indicate the linearity and the slopes of the RCM trajectories, respectively. Since the slopes of the trajectories of the stationary targets and the moving targets are different, the peaks are located at different locations. By differentiating the locations of the peaks, the stationary targets can be suppressed and the along-track moving targets can be detected. The relationship between the angle θ of Hough transform and the along-track velocity is

$$\tan\theta = \frac{\lambda}{cK_r}\frac{2V_aV_y - V_y^2}{R_0}\frac{f_s}{PRF} \tag{5.4.6}$$

The flowchart of the proposed algorithm is shown in Fig. 5.11.

Simulated data is employed to evaluate the performance of the proposed algorithm. Three stationary targets (ST1, ST2, and ST3) and two moving target (MT1 and MT2) are set in the scene, and an alpha-stable distribution noise with stability parameter α equals 1.5 is added to the echo to simulate the clutter of the real FMCW SAR data. The system parameters of the typical FMCW SAR are listed in Table 5.1, and the information of the five targets is listed in Table 5.2.

Figure 5.12a shows the results of the Hough transform after the azimuth down-sampling and the range interpolation. The down-sampling rate and range interpolation rate are set as 4 and 32, respectively. The larger these two parameters are, the more distinctive the residual rang walk migration will be. However, the ratio of the interpolation rate to the down-sampling rate determines the data amount and calculation burden, thus the values should be chosen eclectically [20]. According to Fig. 5.12a, the peaks of the stationary targets are located at regions around $\theta = 0°$ and $\theta = 180°$, and the peaks of the moving targets are located away from these regions. Figure 5.12b illustrates the result after setting the regions θ = 0°–5° and θ = 175°–180° to zeros. The spectra of the stationary targets have been suppressed, and the peaks of the moving targets are contained. The amplitude image of all the targets, which is the indication of the linearity of the trajectories, is shown in Fig. 5.12c. The amplitude peaks (tagged by the rectangles) of ST1, ST2, and ST3 are 9, 11, 7, respectively, and those of MT1 and MT2 are 24 and 21, respectively. It is noted that the amplitudes of the moving targets are larger than those of the stationary targets. The result of the CFAR detection is shown in Fig. 5.12d. We can see from Fig. 5.12d that the energy of the stationary targets is eliminated, and the moving targets MT1 and MT2 are detected.

With the result of the Hough transform, the along-track velocities of the moving targets can also be estimated, and in turn the targets can be focused and relocated in the stationary image. Note that before azimuth refocusing, the additional range walk of the moving target must be corrected. Figure 5.13 shows the comparison of azimuth imaging of MT1 before and after additional range walk correction. From

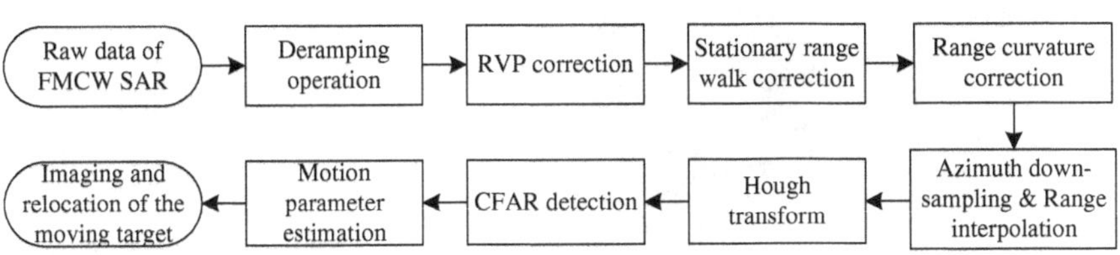

Fig. 5.11 Flowchart of the proposed algorithm

Table 5.1 System parameters

Parameter	Value
Carrier frequency	1×10^{10} Hz
Pulse repetition frequency	500 Hz
Pulse bandwidth	1×10^{8} Hz
Range sampling rate	2.67×10^{5} Hz
Central slant range	1000 m
Reference range	600 m
Antenna size	1 m
Platform velocity	100 m s^{-1}

Table 5.2 Target parameters

Target	Slant range (m)	Along-track velocity (m s^{-1})
MT1	1000	10
MT2	900	20
ST1	1200	0
ST2	1100	0
ST3	800	0

Notes MT1 and MT2 represent moving target 1 and 2, respectively. ST1, ST2, and ST3 represent stationary target 1, 2, and 3, respectively

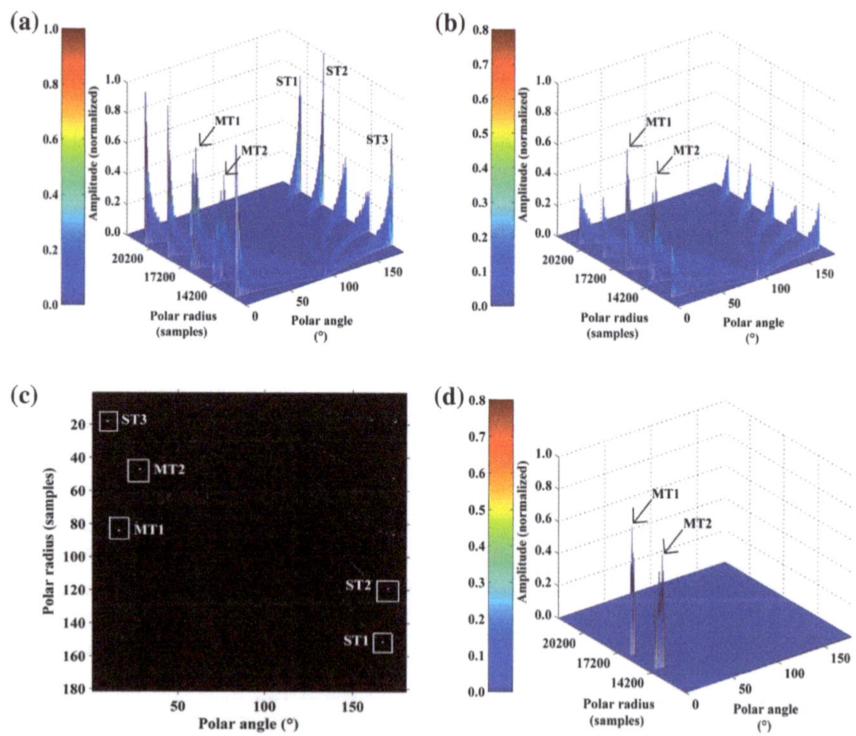

Fig. 5.12 Result of Hough transform, **a** before and **b** after the removal of the stationary targets. **c** The amplitude of peaks of the five targets after the clutter suppression. **d** CFAR detection results

Fig. 5.13 Azimuth compression result for the target MT1

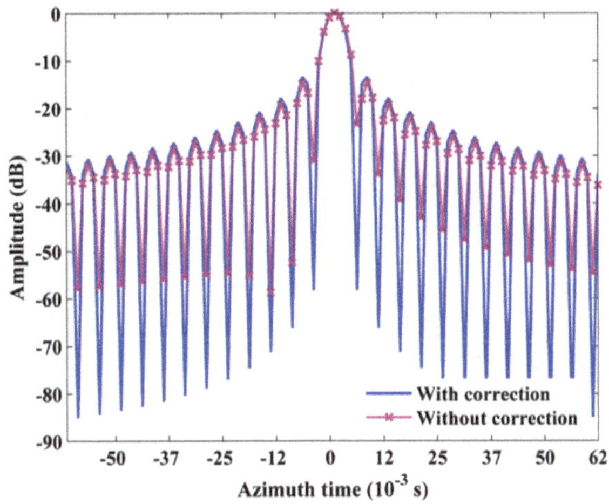

Table 5.3 Imaging quality analysis

Target	Range direction			Azimuth direction		
	IRW (samples)	PSLR (dB)	ISLR (dB)	IRW (samples)	PSLR (dB)	ISLR (dB)
MT1	1.850	−25.320	−2.622	3.000	−14.706	−9.830
ST1	1.800	−25.312	−2.556	3.000	−14.827	−9.928

Notes MT1 represents moving target 1, and ST1 represents stationary target 1. *IRW* impulse response width, *PSLR* peak side lobe ratio, *ISLR* integrated side lobe ratio

Fig. 5.13, it can be observed that the azimuth resolution is improved after the correction of the additional range walk. Detailed imaging quality between MT1 and ST1 is compared in Table 5.3. From Table 5.3, the focusing performance of the moving target is convincing since it shares the same imaging quality with stationary targets.

5.5 Conclusion

The nonlinearity correction and GMTI issues in FMCW SAR have been discussed in this chapter. As a new scheme of SAR, FMCW SAR acquires high range and azimuth resolutions by continuously transmit and receive signals. The small-size, low-cost, and high resolution characters make FMCW SAR highly suitable for UAV platforms, which is a hotspot in the military applications.

The nonlinearity correction is the prior problem to be solved in the FMCW SAR signal processing. The impact of nonlinearity is not negligible since the foundation of the following signal processing operations. Therefore, a novel nonlinearity correction algorithm based on homomorphic deconvolution has been introduced in this chapter. The advantages of this algorithm are: first, this algorithm is a theoretically unbiased algorithm compared with the traditional derivative algorithm; second, this algorithm does not rely on the length of the delay line.

After the nonlinearity correction, the signal model of a moving target in FMCW SAR has been established in Sect. 5.4. By extract the additional RWM with Hough transform, the along-track moving target can be indicated, and the motion parameters can be estimated.

The research on FMCW SAR signal processing in this chapter is still in theory and simulation stage. The algorithms introduced in this chapter are validated by mathematics and simulations, however, real FMCW SAR data are needed to further demonstrate the effectiveness of these algorithms.

References

1. Ribalta A (2011) Time-domain reconstruction algorithms for FMCW-SAR. IEEE Geosci Remote Sens Lett 8(3):396–400
2. Griffiths HD (1988) Synthetic aperture processing for full-deramp radar altimeters. Electron Lett 24(7):371–373
3. Meta A, Hoogeboom P (2003) High resolution airborne FM-CW SAR: digital signal processing aspects. IEEE international geoscience and remote sensing symposium, pp 4074–4076
4. Meta A (2006) Signal processing of FMCW synthetic aperture radar data. Doctoral dissertation, Delft University of Technology
5. Meta A, Hoogeboom P, Ligthart LP (2007) Signal processing for FMCW SAR. IEEE Trans Geosci Remote Sens 45(11):3519–3532
6. Stove AG (1992) Linear FMCW radar techniques. In: IEE proceedings-F, October 1992, vol 139, no 5, pp 343–350
7. Beasley PDL, Stove AG, Reits BJ, et al (1990) Solving the problems of a single antenna frequency modulated CW radar. In: Proceedings of IEEE international radar conference, pp 391–395
8. Liang Y (2009) Signal processing of FMCW SAR. Xidian University, Xi'an
9. Zhang D (2007) Airborne SAR real-time signal processing and FMCW SAR imaging technique. Institute of Electronics, Chinese Academy of Sciences, Beijing
10. Park HG, Kim B, Kim YS (2000) VCO nonlinearity correction scheme for a wideband FM-CW radar. IEEE Microw Opt Technol Lett 25(4):266–269
11. Mizutani H, Tsuru M, Matsuzuka T, et al (2006) A millimeter-wave highly linear VCO MMIC with compact tuning voltage converter. In: Proceedings of EuMC, September 2006, pp 1755–1758
12. Sanchez M, del Casar MA, Briso C, et al (2003) Application of the fractional synthesizers to SFCW radars. In: Proceedings of EuMC, October 2003, pp 1207–1210
13. Jiang W, Pennock SR, Shepherd PR (2012) FMCW radar range performance improvement with modified adaptive sampling method. IEEE Geosci Remote Sens Lett 9(4):668–671
14. Kulpa KS, Wojtkiewicz A, Nalecz M et al (2000) The simple method for analysis of nonlinear frequency distortions in FMCW radar. In: Proceedings of international conference of microwave and radar wireless communication, pp 235–238
15. Meta A, Hoogeboom P, Ligthart LP (2006) Range non-linearities correction in FMCW SAR. In: IEEE international geoscience and remote sensing symposium, pp 403–406

16. Bogert BP, Healy MJR, Turkey JW (1963) The quefrency analysis of time series for echoes: cepstrum, pseudo-autocovariance, cross-cepstrum, and saphe cracking. In: Time series analysis, pp 209–243
17. Zou M (2001) Deconvolution and signal recovery. National Defend Industry Press, Beijing
18. Liang Y, Wang H, Xing M et al (2011) Slow ground moving target parameter estimation and imaging in FMCW SAR. Syst Eng Electr 33(5):1001–1006
19. Meta A, Hoogeboom P (2005) Signal processing algorithms for FMCW moving target indicator synthetic aperture radar. In: IEEE international geoscience and remote sensing symposium, pp 316–319
20. Li Y, Wang Y, Liu C (2011) Detect and focus the moving target by its range walk in time domain. In: International conference on wireless communication and signal processing, November 2011, pp 9–11

Chapter 6
Non-ideal Motion Error Analysis in GMTIm

Abstract This chapter mainly discusses the non-ideal motion error compensation issue in GMTIm with real SAR data. Pointing at the problem that existing GMTIm algorithms have poor performances in the real SAR data processing, the impact of Doppler centroid is analyzed. The signal model of a moving target with non-ideal motion error is established, and the platform velocity error and cross-track velocity error of the moving target are analyzed. Moreover, an error estimation and compensation algorithm is presented, and a whole practical SAR data processing scheme with the algorithms in the frontal chapters is proposed. Finally, simulations and real data are utilized to prove the effectiveness of these algorithms.

6.1 Introduction

The GMTI and GMTIm algorithms in several main working modes are discussed in the frontal chapters. Compared with GMTI, the GMTIm has a higher requirement to the signal processing. GMTIm uses the out-comings of GMTI, and is the foundation of the target recognition.

In traditional GMTIm algorithms, it is assumed that the moving target will be perfected focused with the precise estimations of the motion parameters. In simulations, this assumption is valid. However, in real data processing, we find that this assumption is valid only when the data has a steady echo character and the target has a simple motion. That is, the moving target cannot be perfectly focused in the existence of non-ideal phase errors even though the GMTIm algorithm is performed. Therefore, non-ideal motion error corrections must be added in the GMTIm algorithm.

This chapter mainly focuses on the non-ideal motion error compensation in GMTIm. The WAS mode has a short synthetic aperture time, thus the motion error is negligible. The geometry of FMCW SAR is the same as pulse SAR, so the stripmap SAR is used in this chapter to analyze the non-ideal motion errors. The motion errors are firstly analyzed, and then PGA algorithm is introduced.

J. Yang, *Study on Ground Moving Target Indication and Imaging Technique of Airborne SAR*, Springer Theses, DOI 10.1007/978-981-10-3075-8_6

After PGA algorithm, the higher order phase errors are assumed to be compensated. However, the moving targets are still unfocused in most real data processing occasions. The impact of the non-ideal motion errors on the Doppler centroid is analyzed in this chapter. The signal model of a moving target with non-ideal motion errors is established, and two sorts of motion errors, i.e., platform velocity error and cross-track velocity error, are the causes of the Doppler centroid smear. A practical GMTIm algorithm is proposed to estimate and compensate these errors, and simulations and real data are used to validate the effectiveness of the algorithm.

This chapter is organized as follows. The motion error and PGA algorithm are introduced in Sect. 6.2. The signal model with non-ideal motion errors is established in Sect. 6.3. Section 6.4 analyzes two sorts of motion errors, and a practical GMTIm algorithm is proposed and testified in Sect. 6.5. In Sect. 6.6, we draw the conclusions.

6.2 Motion Error Compensation in Stationary SAR Imaging

6.2.1 Motion Error Analysis in Airborne SAR

In airborne SAR imaging, the platform is assumed to be moving along a straight trajectory with a constant velocity. However, in practical applications, these conditions cannot be satisfied, and the motion error compensation must be induced in the imaging process. The motion error of the stationary target imaging is the same as that of moving target imaging, so the motion error compensation is necessary in GMTIm algorithms.

The geometry of airborne SAR with motion error is illustrated in Fig. 6.1. It can be noted that the actual trajectory of the platform is a non-ideal trajectory. Suppose

Fig. 6.1 Geometry of an airborne SAR with the motion error

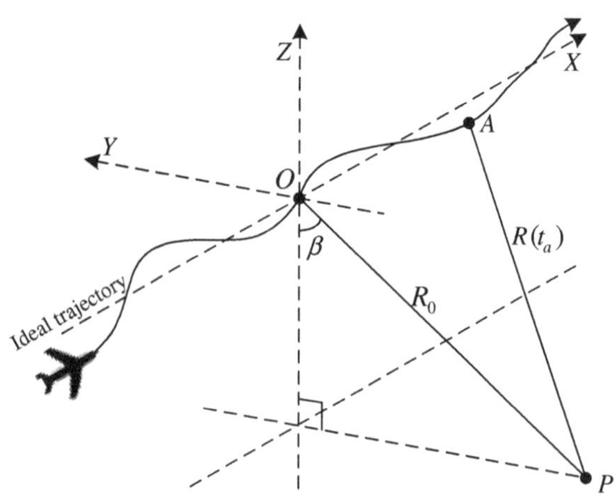

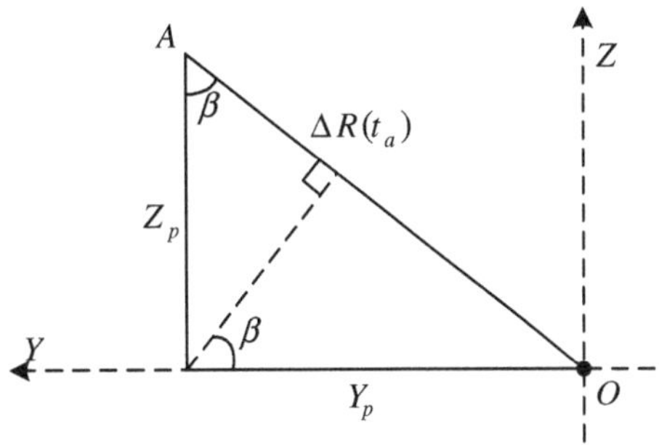

Fig. 6.2 Geometry of the range difference

the coordinate of sample A is $(X(t_a), Y(t_a), Z(t_a))$, the coordinate of Target P is (X_p, Y_p, Z_p), the instantaneous slant range from A to P is

$$R(t_a) = \sqrt{(X(t_a) - X_p)^2 + (Y(t_a) - Y_p)^2 + (Z(t_a) - Z_p)^2} \tag{6.2.1}$$

After Taylor expansion and neglect the higher order phase of t_a, it yields

$$R(t_a) \approx R_0 + \frac{(X(t_a) - X_p)^2}{2R_0} - Y(t_a)\sin\beta - Z(t_a)\cos\beta \tag{6.2.2}$$

where the nearest range of Target P $R_0 = \sqrt{Y_p^2 + Z_p^2}$, $\sin\beta = \frac{Y_p}{R_0}$, $\cos\beta = \frac{Z_p}{R_0}$. The last two terms in Eq. 6.2.2 is the range difference between the actual trajectory and the ideal trajectory, as shown in Fig. 6.2 Suppose the range difference is $\Delta R(t_a)$, i.e., $\Delta R(t_a) = -Y(t_a)\sin\beta - Z(t_a)\cos\beta$, substitute $\Delta R(t_a)$ into Eq. 6.2.2, it yields

$$R(t_a) \approx R_0 + \frac{(X(t_a) - X_p)^2}{2R_0} + \Delta R(t_a) \tag{6.2.3}$$

It can be noted from Eq. 6.2.3 that the motion error of the platform is separated into two parts: the inconstant velocity of the platform and the inconstant slant range. Both errors will cause the error in the instantaneous slant range, and in turn lead to the azimuth smear [1]. Two kinds of errors and the compensation methods are introduced below.

(a) Inconstant velocity error of the platform

In an ideal SAR system, the platform is assumed to be moving with a constant velocity, and the signal is transmitted with PRF. The signal is transmitted and received with a constant PRF, and the azimuth signal is uniformly sampled. However, if the PRF remains constant, and the platform velocity is changing, the azimuth signal sampling is no longer uniform, and the azimuth resolution will be deteriorated. The geometry of azimuth non-uniform sampling is illustrated in Fig. 6.3.

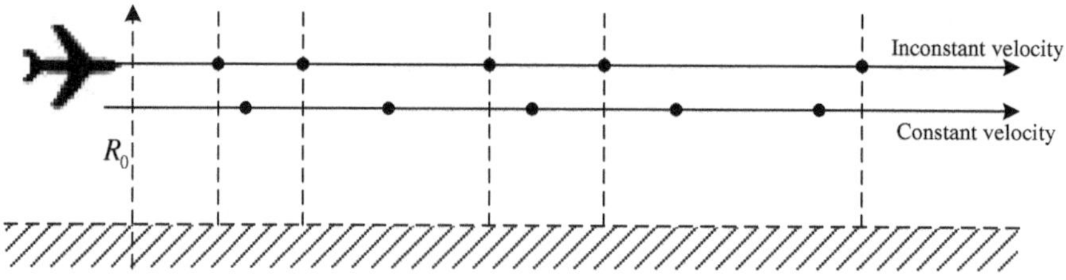

Fig. 6.3 Geometry of the azimuth non-uniform sampling

In a signal processing point of view, according to Eq. 2.2.6, the Doppler modulation rate of a stationary target can be expressed as

$$f_{dr} = \frac{2V_a^2}{\lambda R_0} \tag{6.2.4}$$

If is V_a no longer constant, f_{dr} will be a changing value. Thus, using a constant f_{dr} as a matched filter will lead to the azimuth defocus.

There are two ways to solve this problem: re-sampling algorithm [2] and phase compensation algorithm. Both algorithms use the velocity data recorded by the INS to compensate the motion error. However, these two algorithms are high dependent to the accuracy of the INS data, and the signal processing complexity is increased.

The most practical way to solve this problem in an airborne SAR system is to change the time-constant PRF into space-constant PRF [3]. By altering the value of PRF according to the velocity of the platform, the PRF-to-velocity ratio K_v remains constant. According to Eq. 2.2.5, the azimuth phase of a target in SAR can be expressed as

$$H_a(t_a) = \exp(-j\pi \frac{2V_a^2}{\lambda R_0} t_a^2) \tag{6.2.5}$$

where azimuth time $t_a = -\frac{T_a}{2} : \frac{1}{PRF} : \frac{T_a}{2}$. Substitute this equation into Eq. 6.2.5, it is noted that the Doppler modulation rate is constant with a constant K_v.

(b) Non-ideal Trajectory Error

Because of the airplane control, wind and other non-ideal conditions, the airplane cannot maintain a straight line in practical flights, and the instantaneous slant range will be changed. The changes of instantaneous slant range will induce two effects: first, an additional phase error will exist in the azimuth echo; second, the echo envelop from each azimuth gate will not be aligned. Therefore, phase error compensation and envelop alignment are the two steps in non-ideal trajectory error compensation.

Substitute Eq. 6.2.3 into Eq. 2.2.5, the additional azimuth phase error can be expressed as

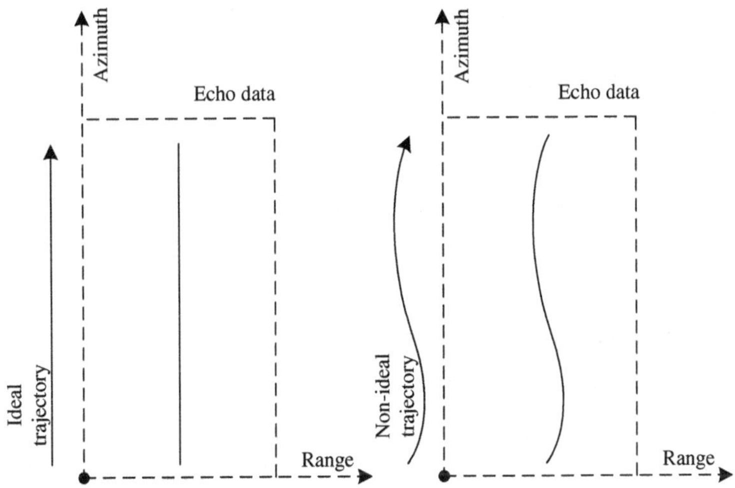

Fig. 6.4 Non-ideal trajectory error

$$H_{\Delta}(t_a) = \exp(-j\frac{4\pi}{\lambda}\Delta R(t_a)) \tag{6.2.6}$$

It can be noted that the phase error can be corrected by multiplying a conjugate filter. The envelop dis-alignment induced by the non-ideal trajectory error is illustrated in Fig. 6.4. Assuming the RCM is corrected, the echo of the target is a straight after the range compression in the ideal condition, while the echo of the target with trajectory error is not straight. Before azimuth compression, the envelop alignment must be operated by data shift or interpolation.

Phase compensation and envelop alignment both need the instantaneous slant range differences on each sampling bin. There are mainly two way to acquire the data: first, use the INS data; second, estimate from the echo data.

6.2.2 *Principle of PGA*

In high resolution imaging of SAR, the error compensation using the INS data is necessary, especially for the airborne SAR. However, constrained by the system cost, the accuracy of the INS data may not satisfy the requirement of error compensation. Therefore, the motion errors must be estimated from the echo. PGA is the most commonly used and effective algorithm in estimating and compensating motion errors from the echo.

PGA algorithm was first used in optical processing, and was introduced in SAR signal processing by Sandia Lab of the U.S. The four practical processing steps of

PGA were presented by D.E. Wahl, which makes PGA algorithm highly suitable for real SAR data processing [4].

The principle of PGA algorithm is: As to a point target after azimuth deramping operation, its azimuth phase only contains the first-order phase that indicates its azimuth location. After correcting this phase, the azimuth phase of the point target will be a constant value. However, if there is phase error, the azimuth phase is no longer a constant value, and the phase error can be estimated by calculating the phase gradient. PGA algorithm uses the strong point target in the scene to estimate the azimuth phase error, and its performance is satisfying in the imaging of areas with multiple strong targets.

The specific steps of PGA algorithm include:

(a) Target Selection and Circular Shifting

The classical PGA algorithm assumes that the azimuth phase error is redundant in the range direction. Therefore the range gates that contain strong point targets have to be selected. The isolated, strong targets are selected, so that the phase error can be accurately estimated. There are different target selection strategies, including Energy Maximum [5], SNR Maximum [6], Contrast Ratio Maximum [7] and so on. The combination of Energy and Contrast Ratio Maximum strategy shows a promising performance in the target selection. Define the contrast ratio C_n of the nth range gate as

$$C_n = \frac{\sqrt{E(I_n^2 - E(I_n^2))^2}}{E(I_n^2)} \tag{6.2.7}$$

where I_n^2 denotes the amplitude of the nth range gate. By setting up the threshold of C_n, the range gate can be selected.

In the selected range gates, the strongest point targets must be circular shifted into the center of the scene to remove the impact of the first-order phase error.

(b) Adding Windows

Adding windows to the strong point target is to contain the phase error and eliminate interference from the phase of other targets and noises. The incorrect length of the windows will induces estimation errors, thus the window must be added accordingly.

In practical PGA algorithms, different window length selection strategies are chosen according to the scene. In the case of isolated strong target, the window is 1.5 multiplies the length of 10 dB [4]. In the case that no isolated strong target exists in the scene, the window is calculated by using the iterations.

An effective window length calculation method is presented in [6]. Suppose the power of an area far from the peak is P_c, the width from the peak to P_c is W_b, the length of fast-descending area is W_a, then the window length can be calculated as

$$W_s = \frac{W_b + W_a}{2} \tag{6.2.8}$$

(c) Phase Gradient Estimation

After circular shift and windowing, perform azimuth IFFT to the signal to estimate the phase gradient. The gradient estimation criterion is selected according to the scene and the SNR of the echo. LUMV algorithm is suitable for the estimation of echoes with high SNR [8]. The core function is

$$\hat{\varphi}'_{LUMV}(t) = \frac{\sum_M \mathrm{Im}[s_w^*(t)s'_w(t)]}{\sum_M |s_w(t)|^2} \tag{6.2.9}$$

Another estimation criterion is Maximum Likeness (ML) estimation [9], with the core function:

$$\Delta\hat{\varphi}_{ML}(n) = \angle[\sum_M s_w^*(n)s_w(n + 1)] \tag{6.2.10}$$

The ML can provide an estimation theoretically equal to the Cramer-Rao bound.

In the images with low SNR, FLOS criterion is often used, which shows a promising performance in the case of heavy-tailed distribution [10]. The core function is

$$\Delta\hat{\varphi}_{FLOS}(n) = \angle[\sum_M s_w(n-1)^{(p_1)} s_w^*(n)^{(p_2)}] \tag{6.2.11}$$

(d) Iterated Phase Correction

Perform integration to the estimated phase gradient, and the estimation of azimuth phase error is obtained. The PGA algorithm is operated iteratively. The threshold of the iteration is selected accordingly.

The classical PGA algorithm is presented in spotlight SAR. In the case of stripmap SAR, several adjustments have to be done.

(a) Cutting the azimuth data into small segments. For each segment, the azimuth phase error is assumed to be invariant. After the estimation of phase gradient in each segment, the phase gradient can be obtained by jointing each segment. The segment cutting is illustrated in Fig. 6.5. It can be noted that each segment is

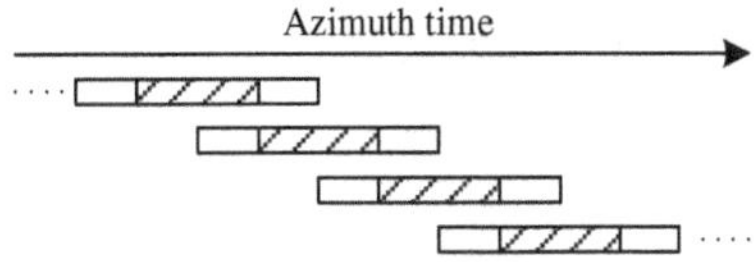

Fig. 6.5 Segments cutting and jointing

50% overlapped with the neighboring segment. By using the central 50% of each segment, the jointing operation can be accurate.

(b) Transform the data from each segment into spotlight data. The spectra of partial observed targets are filtered in this step.

(c) Perform standard PGA algorithm in each segment, and joint the estimations of each segments to obtain the whole estimation. To remove the linear phase error in jointing, the jointing operation uses the second-order phase gradient.

6.3 Signal Model of Moving Target with Non-ideal Motion Error

The high order phase error is assumed to be compensated by PGA and INS data. However, the images of moving targets are mostly smeared in real SAR data processing. This indicates that there are other non-ideal phase errors that lead to the de-focusing.

We believe that the first-order azimuth phase error is the cause of the de-focusing. In stationary SAR imaging, the energy of a target is assumed to be scattered from a single Doppler centroid. In practical SAR systems, the LOS of the antenna remains steady, so that the Doppler centroid of the clutter is single. Therefore, the focused stationary SAR image can be obtained after high order phase error is compensated.

However, in the case of moving targets, the single Doppler centroid may not be satisfied. In our research, we found that the Doppler centroid of a moving target is not a constant value in practical SAR data, and it will severely deteriorate the imaging quality of moving targets.

In traditional GMTI and GMTIm algorithms, the platform is modeled as moving with a constant velocity along a straight course, and the moving target is modeled as moving with a constant velocity and acceleration during the synthetic aperture time. The typical synthetic aperture time of an airborne SAR is around 1–2 s, so the assumption is rational in most cases. However, in the existence of non-ideal motion errors, a more precise signal model must be established.

Since the non-ideal trajectory error can be compensated by INS and PGA, the airplane is assumed to move along a straight trajectory. The platform velocity $V_a(t_a)$ is variant to azimuth time. The geometry of a moving target with non-ideal motion error is illustrated in Fig. 6.6.

In the slant range domain, the cross-track velocity is denoted by $V_r(t_a)$, the along-track velocity is denoted by $V_y(t_a)$. When the platform flies O from to O_1, the moving target moves from P to P_1. Suppose the nearest slant range of the moving target is R_0, the instantaneous slant range $R(t_a)$ can be expressed as

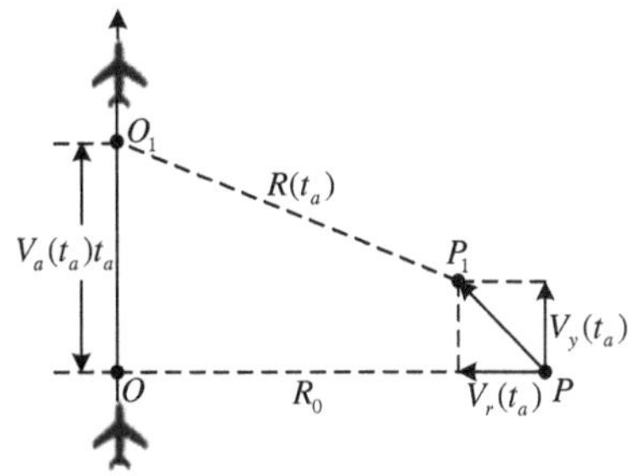

Fig. 6.6 Geometry of a moving target with non-ideal motion errors

$$R(t_a) = \sqrt{(R_0 - V_r(t_a)t_a)^2 + (V_a(t_a) - V_y(t_a))^2 t_a^2} \tag{6.3.1}$$

Expand Eq. 6.3.1 into a Taylor series and keep it to the second-order term of t_a, then Eq. 6.3.1 can be approximated as

$$R(t_a) \approx R_0 - V_r(t_a)t_a + \frac{(V_a(t_a) - V_y(t_a))^2}{2R_0} t_a^2 \tag{6.3.2}$$

Substitute Eq. 6.3.2 into Eq. 2.2.4 and expand, it yields

$$\begin{aligned} s_r(t_a, t_r) &= A_0 \omega_a(t_a) \sin c[\pi B_r(t_r - \frac{2R_0}{c})] \exp(j\frac{4\pi}{\lambda} V_r(t_a)t_a) \\ &\cdot \exp(-j\frac{4\pi}{\lambda} \frac{(V_a(t_a) - V_y(t_a))^2}{2R_0} t_a^2) \exp(j\frac{4\pi K_r}{c} V_r(t_a)t_r t_a) \\ &\cdot \exp(-j\frac{4\pi K_r}{c} \frac{(V_a(t_a) - V_y(t_a))^2}{2R_0} t_r t_a^2) \exp(-j\frac{4\pi}{\lambda} R_0) \end{aligned} \tag{6.3.3}$$

In Eq. 6.3.3, the first two terms are the azimuth phase of the target. The first denotes the Doppler centroid of the target, and the second denotes the second- and higher-order phase error. The third and fourth terms are the RCM of the moving target. In traditional SAR imaging algorithm, the focus of a target mainly relies on the compensation of the second term in Eq. 6.3.3. In the GMTIm, this term can be accurately compensated by the precise motion parameter estimation and PGA.

However, the PGA algorithm cannot correct the error caused by the Doppler centroid. In Eq. 6.3.3, the Doppler centroid is related to t_a and $V_r(t_a)$. If the Doppler centroid is constant, the energy of the target will be concentrated into a point after the higher order azimuth phase error is corrected. However, since t_a and $V_r(t_a)$ are inconstant, the energy of moving target will not be concentrated into a point, which makes the image of the target smear in azimuth.

Azimuth time t_a is sampled by PRF, and PRF is determined by the platform velocity $V_a(t_a)$ in a space-constant PRF system. Therefore, the Doppler centroid is affected by both the platform and the moving target.

6.4 Motion Error Analysis

6.4.1 Platform Velocity Error

In the stationary SAR imaging, the platform velocity error will lead to the inconstant Doppler modulation rate. In order to solve this problem, the SAR system is changed from the time-constant sampling system to the spatial-constant sampling system. The PRF of the spatial-constant sampling system is instantaneously adjusted to the platform velocity, so as to keep the PRF-to-velocity ratio constant. In such systems, the time-variant PRF can be defined as

$$PRF(t_a) = K_v \cdot V_a(t_a) \tag{6.4.1}$$

In the spatial-constant sampling system, the Doppler bandwidth of the clutter can be expressed as

$$B_a(t_a) = \frac{2V_a(t_a)}{L_a} \tag{6.4.2}$$

where L_a denotes the azimuth size of the antenna. According to Eqs. 6.4.1 and 6.4.2, both $B_a(t_a)$ and $PRF(t_a)$ are linear functions of $V_a(t_a)$. Therefore, the spectrum of the clutter in each azimuth sample can be coherently accumulated during T_a. However, the Doppler centroid of the moving target is independent of $V_a(t_a)$. If $V_a(t_a)$ is inconstant, the location of f_{dc} in each azimuth sample is different from one another, which will make the spectrum of the moving target spread on multiple Doppler centroids. The impact of the platform velocity error is illustrated in Fig. 6.7.

In Fig. 6.7, the red breaking lines represent the Doppler histories of the moving target. In Fig. 6.7a, the Doppler history of the moving target is a rectilinear trajectory. On the contrary, the trajectory is not a straight line in Fig. 6.7b with the PRF variation. In the case of Fig. 6.7b, the energy of the moving target cannot be compressed into one Doppler centroid. In practical airborne SAR applications, especially for unmanned missions, the platform velocity changes severely in the existence of the air turbulence. Therefore, the platform velocity error must be compensated in the moving target imaging.

6.4.2 Cross-Track Velocity Error of the Moving Target

In most practical situations, the cross-track velocity of the moving target cannot maintain constant during the entire data acquisition T_a. If the cross-track velocity is inconstant, the energy of the moving target will spread on multiple Doppler centroids.

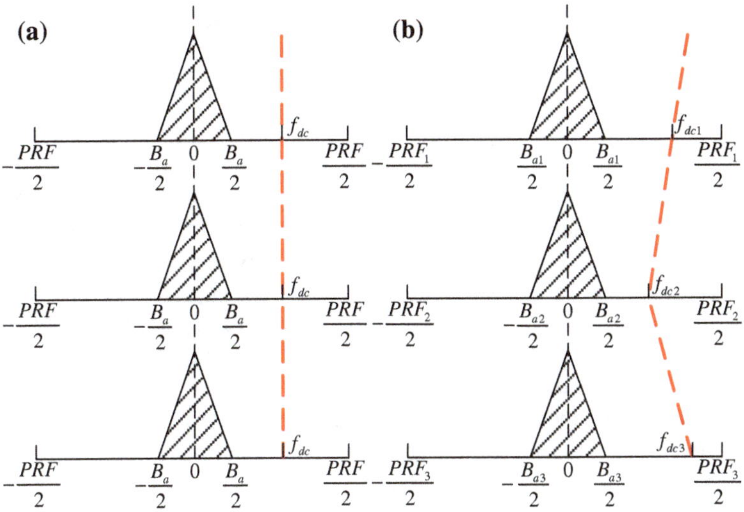

Fig. 6.7 Impact of the platform velocity error on the Doppler centroid of the moving target

The cross-track velocity error has a significant impact on the Doppler centroid of the moving target. For example, consider a typical Ku-band airborne SAR system, in which λ is 0.0194 m, R_0 is 19000 m, and V_a is 100 m/s, the Doppler centroid of the moving target can be calculated as

$$f_{dc} = \frac{2V_x}{\lambda} \approx 103.1 \cdot V_x \tag{6.4.3}$$

Suppose V_x is 10 m/s, then f_{ac} is 1031 Hz. However, if $V_x(t_a)$ varies from 9 to 11 m/s, f_{dc} changes between 927.9 and 1134.1 Hz, as shown in Fig. 6.8. In other words, in the existence of the cross-track velocity error, the moving target cannot be focused into a single Doppler centroid, even though the high-order phase errors are compensated. Therefore, the cross-velocity error of the moving target must be compensated in the real data processing.

Fig. 6.8 Doppler centroid of a target with cross-track velocity error in time-frequency domain

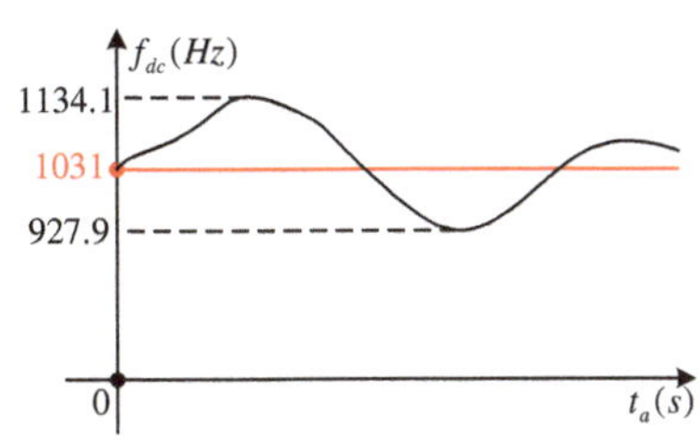

6.5 Non-ideal Motion Error Estimation and Correction Algorithm

6.5.1 Algorithm Principle

To achieve an acceptable imaging quality in real application, the motion error must be estimated and compensated. Two sorts of motion errors can be compensated separately according to the processing steps as follows.

The INS data can be utilized for the platform velocity error compensation. In modern airborne SAR systems, the platform velocity history is recorded by the INS system. With the actual platform velocity $V_a(t_a)$, the actual PRF history $PRF(t_a)$ can be calculated according to Eq. 6.4.1.

As to the single moving target imaging, the compensation of the platform velocity error can be considered as a scaling transformation problem. The aim of the correction is to adjust the PRF in each azimuth sample to the same scale. The correction filter in 2-D time domain can be expressed as

$$H_{Va} = \exp(-j2\pi \frac{\bar{f}_{dc}}{PRF_{ave}}(PRF(t_a) - PRF_{ave})t_a) \tag{6.5.1}$$

where PRF_{ave} represents the average PRF, and $\bar{f}_{dc}$ is the average Doppler centroid of the moving target that estimated by Hough transform.

After the platform velocity error compensation, the platform velocity and the PRF of the system are considered constant, and the residual Doppler centroid error is regarded as the result of the cross-track velocity error.

Since the motion of the moving target is not recorded by the system, the actual motion of the moving target should be estimated from the echo. After the RCM correction, the energy of the moving target is concentrated into a single range cell. By using the FrFT, the azimuth time-frequency distribution of the moving target can be obtained, and the Doppler centroid error caused by the cross-track velocity error can be estimated from it. The principle of our cross-track velocity error estimation algorithm is shown in Fig. 6.9.

A reference moving target without motion errors can be simulated, and the time-frequency distribution of the ideal moving target can be used to extract the information of the cross-track velocity error. The Doppler centroid error $\Delta\hat{f}_{dc}$ can be extracted from the time-frequency distribution of the moving target, and the cross-track velocity error can be obtained as

$$\Delta\hat{V}_r(t_a) = \frac{\Delta\hat{f}_{dc} \cdot \lambda}{2} \tag{6.5.2}$$

The correction filter of the cross-track velocity error can be expressed as

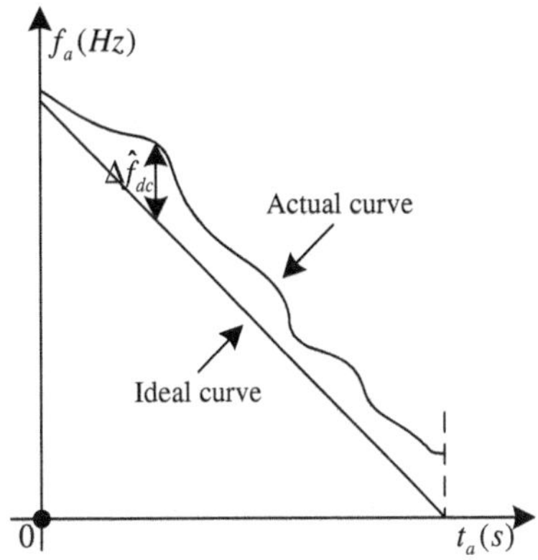

Fig. 6.9 The principle of the motion error estimation algorithm

$$H_{Vr} = \exp(-j\frac{4\pi}{\lambda}\Delta V_r(t_a)t_a) \tag{6.5.3}$$

After the motion error compensation, the imaging resolution of the moving target can be significantly improved. Using the CLEAN technique [11], the flowchart of our whole moving target processing strategy can be illustrated as shown in Fig. 6.10.

6.5.2 *Simulation and Raw Data Processing*

In the simulation, one moving target with inconstant cross-track velocity is set at the center of the scene according to the system parameters listed in Table 3.1. The average cross-track velocity of the moving target is 10 m/s, whereas its actual cross-track velocity varies during T_a as shown in Fig. 6.11a. The platform velocity is not constant during T_a either, and it is shown in Fig. 6.11b. The proposed motion error compensation algorithm is implemented to estimate and correct the motion error of the moving target.

The platform velocity error is firstly compensated with INS data. According to our cross-track velocity error estimation algorithm, a reference signal with the cross-track velocity of 10 m/s is simulated, and the time-frequency distributions of the moving target and the reference target are shown in Fig. 6.11c. Together with Fig. 6.11a, it can be noted that the time-frequency distribution accurately reflects the changes of the cross-track velocity. The estimated velocity error is shown in Fig. 6.11d. We can see that the actual cross-track velocity of the moving target is correctly estimated by our algorithm. The imaging results of the moving target before and after the compensation are compared in Fig. 6.11e, f. The image of the moving target splits into several peaks as shown in Fig. 6.11e, which indicates that

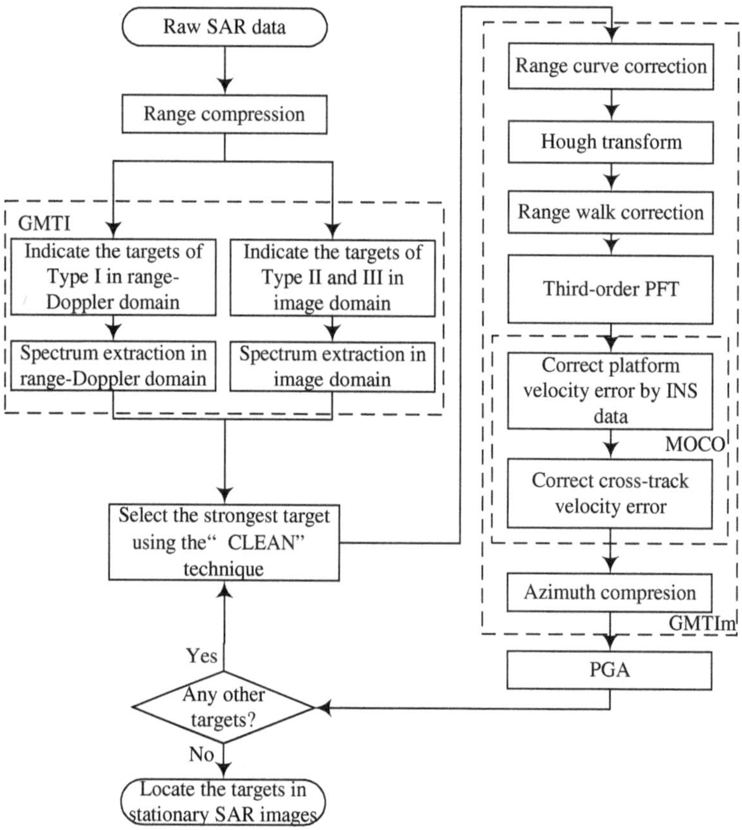

Fig. 6.10 Flowchart of the proposed moving target processing strategy. MOCO represents motion error compensation

the moving target has more than one Doppler centroid. After the compensation, the imaging resolution is significantly improved, as shown in Fig. 6.11f.

The same Ku-band real airborne SAR data is used to prove the effectiveness of this algorithm. Here, we use T2 as an example to testify the performance of the proposed motion error compensation algorithm in the real data processing.

The time-frequency distributions of the moving target and the reference target are shown in Fig. 6.12a. The imaging results before and after the compensation are compared in Fig. 6.12b. We can see from Fig. 6.12b that the imaging resolution is significantly improved after the compensation.

The final imaging result of the seven moving targets with the proposed moving target processing strategy is shown in Fig. 6.13. All the moving targets (tagged by arrows and rectangles) are detected by the proposed strategy, and they are also focused and relocated to their actual locations. Therefore, the effectiveness of the proposed strategy in the real data processing is confirmed.

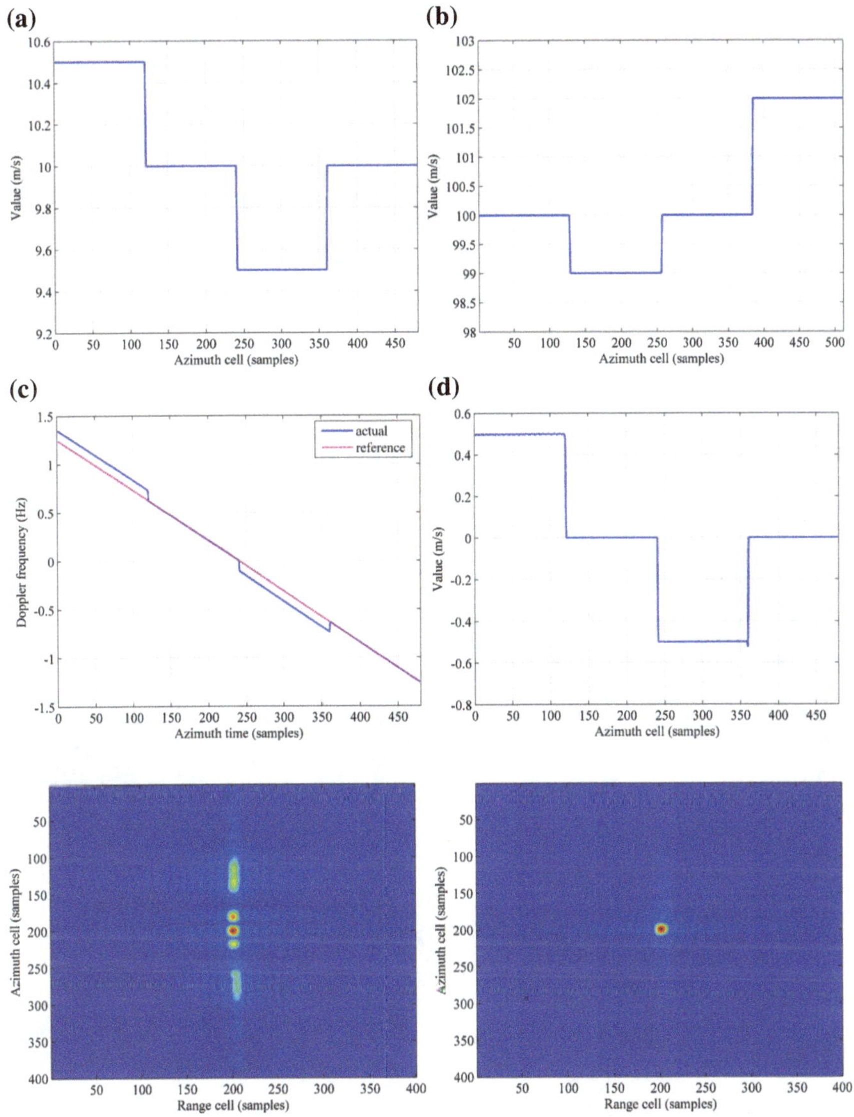

Fig. 6.11 The processing results of motion error compensation. **a** Actual cross-track velocity, **b** Actual platform velocity, **c** Time-frequency distribution of the moving target, **d** Result of the cross-track velocity estimation, **e** Imaging before the motion error compensation, **f** Imaging after the motion error compensation

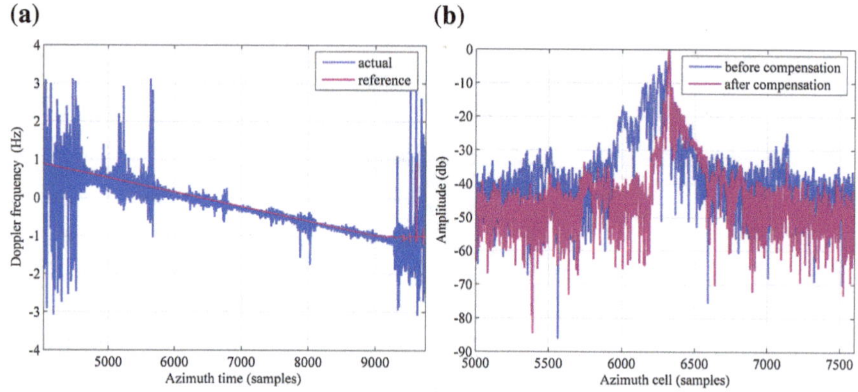

Fig. 6.12 Results of the motion error compensation. **a** Time-frequency distributions of T2 and the reference signal. **b** Imaging comparison of T2 before and after the motion error compensation

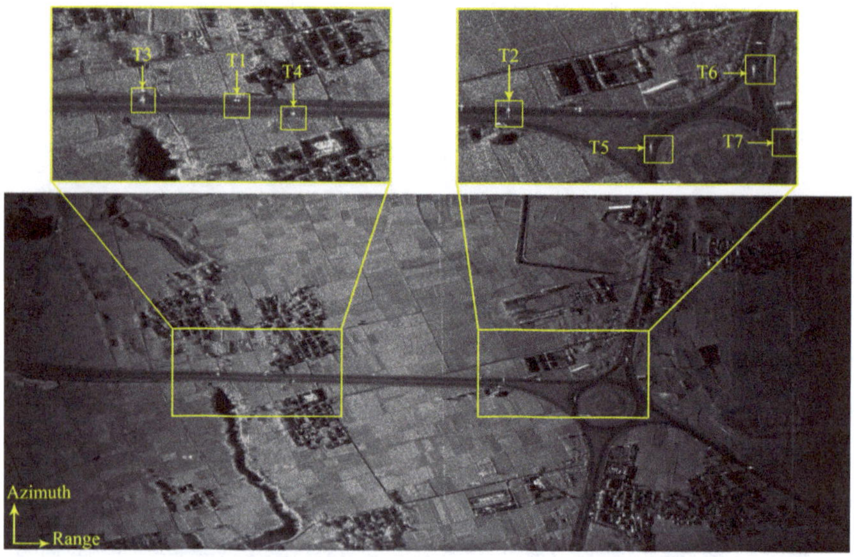

Fig. 6.13 Imaging of the moving targets in the stationary image with the proposed strategy

6.6 Conclusion

This chapter mainly focuses on the non-ideal motion error estimation and compensation in real SAR data GMTIm. The existing GMTIm algorithms have been proved valid and effective in simulations, but in real data processing, their performances are degraded. It indicates that there are non-ideal motion errors that must be compensated in the real data processing.

Two sorts of non-ideal motion errors have been analyzed, and the estimation and compensation algorithms have been presented. Simulations and real data have been utilized to prove the effectiveness of the algorithms.

The main contribution of this chapter is to analyze the impact of the Doppler centroid error on the imaging of moving targets. This research is still on the exploratory stage, while it appears to be inspirational in the GMTIm with real SAR data. Along with following researches, there may be other non-ideal errors, and the GMTIm of real SAR data will be further improved.

References

1. Li Y, Liu C, Wang Y et al (2012) A robust motion error estimation method based on raw SAR data. IEEE Trans Geosci Remote Sens 50(7):2780–2790
2. Chen L, Liang X, Ding C (2010) Non-uniform reconstruction method in SAR imaging. J Syst Simul 22(5):1242–1245
3. Zhang Z (2003) Introduction to airborne and spaceborne synthetic aperture radar. Publishing House of Electronics Industry, Beijing
4. Wahl DE, Eichel P, Ghiglia DC et al (1994) Phase gradient autofocus—a robust tool for high resolution SAR phase correction. IEEE Trans Aerosp Electron Syst 30(3):827–835
5. Grewal MS, Weill LR, Andrews AP (2007) Global positioning systems, inertial navigation, and integration. Wiley, Hoboken
6. Chen Q, Li J (2004) Performance analysis and improvement of phase gradient autofocus algorithm. J Beijing Univ Aeronaut Astronaut 30(2):131–134
7. Zhao X, Wang X, Wang Z (2005) Phase gradient autofocus algorithm for SAR imagery based on contrast criteria. Remote Sens Technol Appl 20(6):606–610
8. Eichel PH, Jakowatz CV Jr (1989) Phase-gradient algorithm as an optimal estimator of the phase derivative. Opt Lett 14(20):1101–1103
9. Jakowatz CV Jr, Wahl DE (1993) Eigenvector method for maximum-likelihood estimation of phase errors in synthetic-aperture-radar imagery. J Opt Soc Am 10(12):2539–2546
10. Tsakalides P, Nikias CL (2011) High resolution autofocus techniques for SAR imaging based on fractional low-order statistics. In: IEE proceedings-radar, sonar and navigation, October 2011, vol 148, no 5, pp 267–276
11. Tsao J, Stenberg BD (1988) Reduction of sidelobe and speckle artifacts in microwave imaging: the CLEAN technique. IEEE Trans Antennas Propag 36(4):543–556

Kolibri 360 GmbH
Ferdinand-Jühlke-Straße 7
99095 Erfurt, Deutschland
produktsicherheit@kolibri360.de